From Model-based Experimental Design and Analysis of Diffusion and Liquid-Liquid Equilibria to Process Applications

Von der modellgestützten experimentellen Analyse von Diffusion und Flüssig-Flüssig-Gleichgewichten zu Prozessanwendungen

Von der Fakultät für Maschinenwesen der
Rheinisch-Westfälischen Technischen Hochschule Aachen
zur Erlangung des akademischen Grades eines Doktors
der Ingenieurwissenschaften genehmigte Dissertation

vorgelegt von

Ludger Wolfgang Michael Wolff

Berichter: Univ.-Prof. Dr.-Ing. André Bardow
Univ.-Prof. Dr.-Ing. habil. Andreas Paul Fröba

Tag der mündlichen Prüfung: 16.03.2020

Aachener Beiträge zur Technischen Thermodynamik Band 30

Ludger Wolfgang Michael Wolff

From Model-based Experimental Design and Analysis of Diffusion and Liquid-Liquid Equilibria to Process Applications
Von der modellgestützten experimentellen Analyse von Diffusion und Flüssig-Flüssig-Gleichgewichten zu Prozessanwendungen

ISBN: 978-3-95886-402-3

Bibliografische Information der Deutschen Bibliothek
Die Deutsche Bibliothek verzeichnet diese Publikation in der Deutschen Nationalbibliografie; detaillierte bibliografische Daten sind im Internet über http://dnb.ddb.de abrufbar.

Herstellung & Vertrieb:

1. Auflage 2021

Süsterfeldstr. 83, 52072 Aachen
Tel. 0241 / 87 34 34 00
www.Verlag-Mainz.de

ISSN: 2198-4832

Satz: nach Druckvorlage des Autors
Umschlaggestaltung: Druckerei Mainz

printed in Germany
D82 (Diss. RWTH Aachen University, 2020)

Danksagung

Die vorliegende Arbeit entstand während meiner Zeit als wissenschaftlicher Mitarbeiter am Lehrstuhl für Technische Thermodynamik der RWTH Aachen University. Sie ist nur durch die Zusammenarbeit mit vielen anderen Personen möglich geworden, denen ich an dieser Stelle danken möchte.

An erster Stelle möchte ich meinem Doktorvater Prof. André Bardow für sein Vertrauen und seine stete Unterstützung danken. Seine Offenheit und sein Anspruch waren immer wieder Motivation und Ansporn zugleich und haben so wesentlich zum Gelingen dieser Arbeit beigetragen. Weiterhin danke ich Prof. Andreas Fröba für die außerordentlich fruchtbare Zusammenarbeit sowie die Übernahme des Koreferats. Auch Prof. Thijs Vlugt möchte ich herzlich für die gute Zusammenarbeit danken. Prof. Christoph Broeckmann danke ich für die Übernahme des Prüfungsvorsitzes und die angenehme Leitung der Prüfung in der turbulenten Zeit der beginnenden Corona-Pandemie.

Einen besonderen Dank möchte ich an alle aktiven und ehemaligen Mitarbeiter des Lehrstuhls für Technische Thermodynamik richten: Es war eine tolle, prägende Zeit! Peter Beumers, Dominique Dechambre, Julia Thien, Christine Peters: Habt Dank für die tolle gemeinsame Zeit, die hervorragende Zusammenarbeit, die langen Diskussionen und die Feierabendbierchen. Herzlichen Dank auch meinen externen Kooperationspartnern Seyed Jamali und Pouria Zangi für die ausgezeichnete Zusammenarbeit, die die Arbeit besonders interessant gemacht hat. Meinen ehemaligen HiWis Daniel Jungen und Mario Echelmeyer danke ich für ihre langjährige Treue und Unterstützung. Thorsten Brands und Hans-Jürgen Koß danke ich für die herzliche Unterstützung in allen fachlichen und organisatorischen Fragen. Einen Dank möchte ich auch allen TaMis aussprechen, insbesondere Iris Wallraven, Katrin Rossbruch, Eva Frach und Gregor Migas in Sekretariat, Buchhaltung und IT, ohne die diese Arbeit ebenfalls nicht möglich gewesen wäre.

Nicht zuletzt möchte ich meinen Eltern, meiner Familie und meinen Freunden danken – für ihren Beistand, ihre Unterstützung und das Leben jenseits der Promotion. Danke für alles!

Aachen, im März 2020 *Ludger Wolff*

Contents

List of Figures

List of Tables

Notation

Abbreviations

MS	Maxwell-Stefan
UNIQUAC	Universal quasi-chemical model
RMSE	root-mean-square error
LJ	Lennard-Jones

Symbols

a_i	weighting factor for reference velocities u
a_{P}	specific mass transfer area of column packing
$A_{\mathrm{R,i}}$	first refractivity virial coefficient
A	mass transfer area
$\boldsymbol{A}_\mu$	derivative of the model equations with respect to the model variables
$\boldsymbol{B}_\mu$	derivative of the model equations with respect to the model parameters
$\boldsymbol{B}$	diffusion coefficient matrix
	for the conversion of Fick and MS diffusion coefficients
B_{ij}	second virial coefficient
BC	base costs
C_{ijk}	third virial coefficient
c_i	molar concentration of species i
c_{t}	total molar concentration
$\bar{c}_i$	dimensionless molar concentration of species i
$\bar{c}_{\mathrm{av},i}$	dimensionless average molar concentration of species i
C_{disp}	disposal costs
C_0	base investment costs
CC_{ij}	cross-correlation function
ccf	capital charge factor
$\boldsymbol{d}$	experimental design vector
$\boldsymbol{d}^*$	experimental design vector of a continuous design
$\boldsymbol{d}^{(N)}$	experimental design vector of an exact design with N experiments

d_{K}	column diameter
$d_{\mathrm{K},0}$	reference column diameter
D	diffusion coefficient
$Đ$	MS diffusion coefficient
$Đ_{\mathrm{Darken}}$	Darken diffusion coefficient
$Đ_{\mathrm{Cross}}$	non-ideal diffusion coefficient
$Đ_{ij}^{x_k \to 1}$	diffusion coefficient at infinite dilution
$D_{i,\mathrm{self}}$	self-diffusion coefficient of species i
$\Delta D_{i,\mathrm{self,rel}}$	relative deviation of predicted self-diffusion coefficients
D^{M}	diffusion coefficient in the molar reference frame
D^{V}	diffusion coefficient in the volume reference frame
$\boldsymbol{F}$	Fisher information matrix
$F_{1-\alpha}$	F-distribution for statistical significance α
F_{m}	material factor
F_{p}	pressure factor
Fo	Fourier number
G	geometry
g^{E}	excess enthalpy
$\boldsymbol{g}$	model equations
h_{K}	column height
$h_{\mathrm{K},0}$	reference column height
H	height
HTU	height of a transfer unit
J^{a}	diffusive flux in reference frame a
J^{M}	diffusive flux in the molar reference frame
J^{V}	diffusive flux in the volume reference frame
k	interference fringe order
k_{diff}	diffusion-induced interference fringe order
$k_{0,\mathrm{i}}$	initial interference fringe order
$k_{\mathrm{drift,i}}$	linear drift parameters for interference fringe orders
l	depth of Loschmidt cell
L	length
ΔL_{opt}	difference in optical path length
$\boldsymbol{M}$	local Fisher information matrix
M_i	molar mass of species i
MF	module factor
MPF	material and pressure factor

m_i	mass of LJ particle of species i
n_i	numbers of moles of species i
n	refractive index
n_{exp}	number of distinct experiments
N	molar flux
N_{c}	number of components
N_{d}	number of measured datapoints
N_{g}	number of model equations
N_{w}	number of model variables
N_{p}	number of model parameters
N_{x}	number of independent model variables
N_{y}	number of dependent model variables
N_{exp}	number of experiments
NTU	number of transfer units
ΔOPL	change in optical path length
p	pressure
Pe	Péclet number
q	interest rate
$\boldsymbol{r}$	position vector
R	gas constant
$\Re$	Reynolds number
s	length coordinate
t	time
t_{a}	annual operation time
T	temperature
TAC	total annual costs
$TAOC$	total annual operational costs
$TUBMC$	total updated bare module costs
u	velocity
u^{a}	reference velocity in reference frame a
u^{M}	molar reference velocity in reference frame a
u^{V}	volume reference velocity in reference frame a
UF	update factor
v_i	design weight of experiment i
$\boldsymbol{v}$	velocity
V	molar volume
V	molar volume of pure species i
$\bar{V}_i^0$	partial molar volume of species i

$\boldsymbol{V}_{\boldsymbol{\theta}}$	
$\boldsymbol{V}_{\boldsymbol{w}}$	measurement (co-)variance matrix
$\boldsymbol{V}_{\boldsymbol{y}}$	measurement (co-)variance matrix of the dependent variables
$\boldsymbol{V}_{\boldsymbol{y},\mathrm{Pred}}$	(co-)variance matrix of the model predictions
$\hat{\boldsymbol{V}}$	(co-)variance matrix of the residuals
$\boldsymbol{w}$	model variables
w_i	mass fraction of component i
W	width
$\boldsymbol{x}$	independent model variables
x_i	mole fraction of component i
X	molar loading
$\boldsymbol{y}$	dependent model variables
y	width coordinate
z	length
z_i	overall mole fraction of component i
z_0	position of the initial domain boundary
$\Delta z_{\mathrm{camera}}$	camera offset
Δz_{fork}	distance of the fork's tines
$\Delta z_{\mathrm{fork,camera}}$	calculated distance of the fork's tines

Greek symbols

β_{ij}	mass transfer coefficient
γ_i	activity coefficient of species i
Γ	thermodynamic factor
δ	thickness of boundary layer
δ_{ij}	Kronecker delta
ϵ_μ	measurement error
ϵ	void fraction of packing
ζ_i	design efficiency
ζ_{D}	D-efficiency of a design
ζ_{G}	G-efficiency of a design
ζ_{c}	c-efficiency of a design
η	viscosity
$\boldsymbol{\theta}$	model parameters
ϑ	temperature
λ	laser wavelength
μ	viscosity
μ_i	chemical potential of species i
ξ	dimensionless length
ξ_0	dimensionless position of the initial domain boundary
ξ_i^{M}	molar concentration of species i in transformed molar reference frame
ρ_{ij}	correlation coefficient
σ^2_{TAC}	variance of a cost funtion TAC
$\tilde{\sigma}_{\boldsymbol{y},\mathrm{Pred}}$	standardized prediction variance of the model predictions
σ_I	interfacial tension
$\boldsymbol{\Sigma}_0$	initial (co-)variance matrix of the model parameters
Φ	objective function
Φ_{D}	objective function for a D-optimal design
Φ_{G}	objective function for a G-optimal design
Φ_{c}	objective function for a c-optimal design
Φ_{LS}	least-squares objective function
Φ_{WLS}	weighted least-squares objective function
Ω	design space

Kurzfassung

Diffusion und Flüssig-flüssig-Gleichgewichte (LLE) sind von entscheidender Bedeutung für das Design verfahrenstechnischer Prozesse wie z.B. der Extraktion. Leider sind bestehende prädiktive Modelle für Diffusionskoeffizienten und LLE häufig nicht hinreichend genau und die Entwicklung neuer prädiktiver Modelle wird durch unverstandene komplexe molekulare Wechselwirkungen und einen Mangel an experimentellen Daten erschwert. Grund für den Mangel an experimentellen Daten ist der hohe Aufwand zur Durchführung von Experimenten.

Das Ziel dieser Arbeit ist die Entwicklung verbesserter Vorhersagemethoden für Diffusionskoeffizienten und die Entwicklung verbesserter Diffusions- und LLE-Messverfahren. Dazu wird ein modellgestütztes Analyseverfahren mit Computerexperimenten und der Entwicklung neuer Messverfahren kombiniert. Die Arbeit ist in enger Zusammenarbeit der TU Delft, der FAU Erlangen-Nürnberg, und der RWTH Aachen University entstanden.

Die Integration des modellgestützen Analyseverfahrens in die Entwicklung der neuen Messverfahren erlaubt eine genaue Analyse der Qualität der experimentellen Ergebnisse. Dadurch werden notwendige Verbesserungen in Modell und Experiment in einem iterativen Verfahren identifiziert und umgesetzt. Das Ergebnis dieses Vorgehens sind neue Diffusionsmessverfahren für Gase und Flüssigkeiten, die eine erhebliche Reduktion des experimentellen Aufwands erlauben.

Zur weiteren Reduktion des experimentellen Aufwands werden in dieser Arbeit optimale Versuchspläne für Diffusions- und LLE-Messungen identifiziert. Optimale Versuchspläne ermöglichen eine Reduktion der *Anzahl* an Experimenten, ohne an Genauigkeit in den Ergebnissen zu verlieren. Wir identifizieren in dieser Arbeit sowohl optimale Versuchspläne für separate Diffusions- und LLE-Experimente als auch optimale Kombinationen von Diffusions- und LLE-Experimenten zur wirtschaftlichen Auslegung von Extraktionsprozessen.

Abstract

Diffusion and liquid-liquid equilibria (LLE) data are of major importance for the design of chemical processes as, e.g., extraction. Unfortunately, predictive equations for diffusion and LLE are not yet sufficiently accurate. The development of predictive equations suffers both from complex molecular interactions that are not fully understood and from a lack of experimental data that would be needed for validation. The reason for the lack of experimental data is that diffusion and LLE measurements are burdened with large experimental effort and high sample consumption.

In this work, we address the need for improved diffusion predictions and experiments as well as for improved LLE experiments. For this purpose, a model-based approach is combined with computer experiments and the development of new measurement methods. The work was conducted in strong collaborations with TU Delft, FAU Erlangen-Nürnberg, and RWTH Aachen University.

The integration of the model-based approach into the development of measurement methods allows for the thorough evaluation of the quality of experimental results. Thereby, necessary improvements in model and experiment are identified and implemented. As a result, effective and consistent diffusion experiments for liquids and gases are developed. The new experiments allow for a significant reduction of measurement times and sample consumption by an order of magnitude.

In addition, optimal experimental designs (OED) for diffusion and LLE experiments are investigated. OED identifies experiments that provide most information on the property of interest. Thereby, the *number* of experiments can be reduced without losing accuracy. We identify OEDs for diffusion and LLE experiments that can reduce the experimental effort by an order of magnitude. In addition, we identify optimal combinations of diffusion and LLE experiments for the economic design of extraction processes.

Thus, the contribution of this thesis is two-fold: On the one hand, efficient experimental designs and setups are developed that allow for a significant reduction of experimental effort. On the other hand, the demonstration of the successful application of the model-based approach can serve as a motivation for an increased integration of models and experimental setups in future developments of experiments.

Chapter 1

Introduction

Physical properties are at the core of any quantitative calculation in chemical engineering. Therefore, accurate knowledge of physical properties is indispensable. Two important classes of physical properties are equilibrium properties and transport properties. Equilibrium properties define the *boundaries* of any thermodynamic process; transport properties describe the *speed* of a thermodynamic process. Thus, the interaction of equilibrium and transport properties defines the progress of a thermodynamic process.

Liquid-liquid equilibria (LLE) and diffusion processes are two important representatives of equilibrium and transport properties. LLE determine, e. g., the distribution of chemicals in extraction and heterogeneous azeotropic distillation processes or in the environment (Gmehling et al., 2012; Michelsen and Mollerup, 2007). Diffusion is an important mass transport process on the micro-scale that is the rate-limiting step in many applications when chemical reactions or heat transfer occur simultaneously (Bird et al., 2002).

Although predictions of LLE and diffusion are improving (Case et al., 2011; Moultos et al., 2016; Sharipov and Benites, 2017), experiments of LLE and diffusion are still indispensable (Gmehling et al., 2012; Economou et al., 2014; Bird, 2004). However, there is a lack of LLE and diffusion data - in particular of multicomponent systems (Economou et al., 2014; Bird, 2004). The reason for the lack of data is that LLE and diffusion experiments are time-consuming and laborious. Therefore, efficiently designed experiments are needed.

The development of efficient experiments requires three important aspects:

1. The experimental setup should be designed in a way that little time and small amounts of samples are consumed,
2. experiments should be chosen such that maximum information on the property of interest can be extracted from the experimental data,
3. the experimental data evaluation has to be carried out in a way that results are statistically significant.

Thus, experiments can be improved from two sides: On the one hand, improved experimental setups can be developed, i.e. the "hardware" can be improved. On the other hand, statistically consistent model-based experimental designs and analyzes can be developed, i.e. the "software" can be optimized. Improvements in the "hardware" correspond to improved measurement techniques whereas improvements in "software" correspond to a model-based approach.

In this work, a model-based approach is applied to design and analyze diffusion and LLE experiments. In Chapter 2, the state of the art of measuring and modeling diffusion and LLE is outlined and the main goals of this work are derived. The method of model-based experimental design and analysis, which builds the foundation for the analyzes of this work, is described in Chapter 3.

In Chapter 4, computer experiments of binary diffusion in non-ideal liquids are analyzed. A new predictive model for concentration-dependent self-diffusion coefficients in non-ideal liquids is developed. The new model provides a missing link for the prediction of binary diffusion. It enables fast calculations and is therefore suitable for implementations in, e.g., process simulations.

In Chapter 5, model-based experimental design techniques are applied to identify optimal geometries for diffusion experiments. Both already existing and theoretically conceivable geometries are considered. Thereby, optimal directions for future developments of diffusion experiments are identified.

In Chapter 6, model-based experimental design techniques are applied for the development of improved diffusion experiments. In Section 6.1, the model-based experimental approach is applied to overcome long-standing discrepancies in the measurement of diffusion in gases with a Loschmidt cell. Thereby, a consistent experimental setup for the measurement of diffusion in gases is developed that enables the measurement of concentration-dependent diffusion coefficients from a single experiment. In Section 6.2, a microfluidic setup for the measurement of diffusion in liquids is developed that enables the measurement of multicomponent diffusion coefficients from a single experiment. Thus, the experimental setups developed with the model-based approach enable significant savings in experimental effort.

In Chapter 7, optimal experimental designs for LLE measurements are identified. These optimal experimental designs can significantly reduce the number of LLE experiments without losing accuracy of the LLE model parameters and predictions.

In Chapter 8, the method of model-based optimal experimental design is applied to an industrial-like extraction process where both diffusion and LLE occur simultaneously. For industrial applications, accurate physical properties are no longer of primary interest, but a minimization of the uncertainty in total annual costs is the main objective. Thus, optimal experimental designs are identified that most reduce uncertainty in costs.

A summary of the results and final conclusions are given in Chapter 9.

Chapter 2

State of the art of diffusion and liquid-liquid equilibria theory, experiments, and predictions

This chapter gives an overview on the state of the art of the measurement and modeling of diffusion and LLE. In Section 2.1, measurement techniques, physical models and predictive engineering models for diffusion are discussed. In Section 2.2, the measurement and modeling of LLE is described. In Section 2.3, the state of the art is discussed and the contribution of this thesis is introduced.

Parts of this chapter are reprinted (adapted) with permission from
Wolff, L., Koß,H.-J., and Bardow, A. (2016). The optimal diffusion experiment, *Chem. Eng. Sci.*, 152:392-402. (Wolff et al., 2016)
Contribution of the author: Writing the draft, principal author, modeling, simulation, analysis.

Peters, C., Wolff, L., Vlugt, T. J. H., and Bardow, A. (2016). Chapter 5 diffusion in liquids: Experiments, molecular dynamics, and engineering models. In *Experimental Thermodynamics Volume X: Non-equilibrium Thermodynamics with Applications*, pages 78-104. The Royal Society of Chemistry. (Peters et al., 2016) - Reproduced by permission of The Royal Society of Chemistry
Contribution of the author: Co-author in writing the draft and final text.

Wolff, L., Jamali, S. H., Becker, T. M., Moultos, O. A., Vlugt, T. J. H., and Bardow,

A. (2018). Prediction of composition-dependent self-diffusion coefficients in binary liquid mixtures: The missing link for Darken-based models, *Ind. Eng. Chem. Res*, 57:14784-14794. (Wolff et al., 2018a), Copyright (2018) American Chemical Society, http://pubs.acs.org/articlesonrequest/AOR-NSu7bFvBu4NFtmTkS8R3,
Contribution of the author: Writing the draft, principal author, analysis of the raw data derived from MD simulations conducted by S. H. Jamali, development of the diffusion model.

2.1 Diffusion

Diffusion is an important mass transfer mechanism in many technical and environmental processes. Scientist and engineers are in constant need of accurate diffusion calculations (Economou et al., 2014). The broad area of applications ranges from the design of industrial processes such as gas separation processes Cussler (2007) to the calculation of geological processes Zhang (2010) such as volcanic eruptions Amalberti et al. (2018a). In recent years, the qualitative and quantitative understanding of diffusion in gases and liquids has substantially improved (Peters et al., 2016; Liu et al., 2011d; Moultos et al., 2016). In this development, non-equilibrium thermodynamics has played a key role. In the following, we summarize the state of the art of measuring, modeling, and predicting diffusion.

2.1.1 Measuring diffusion

Diffusion experiments are the most significant source for diffusion coefficients (Peters et al., 2016). Diffusion measurements are typically executed by the following procedure (cf. Figure 2.1(a)):

1. Two domains are put into contact that contain mixtures with a difference in concentrations $\boldsymbol{c}_0$ and $\boldsymbol{c}_1$.
2. After the start of the diffusion process, concentrations are measured by an appropriate analytical technique as a function of space and/or time. Important analytical techniques for diffusion experiments are, e.g., interferometry, refractive index, Raman-spectroscopy, chromatography, and conductivity (Peters et al., 2016; Woolf et al., 1991).

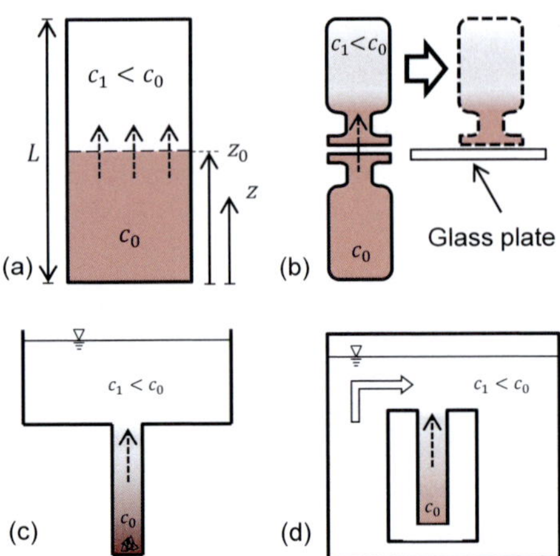

Figure 2.1: Diffusion geometries with concentration gradients indicated by dashed arrows. a) Typical initial conditions for a closed diffusion geometry. b) Graham's diffusion experiment with a closed geometry (redrawn after Cussler (2007)). c) Fick's diffusion experiment with an open geometry and time-independent concentration profile (redrawn after Cussler (2007)). d) Anderson's and Saddington's diffusion experiment with a capillary as open geometry (redrawn after Woolf et al. (1991)).

3. The diffusion coefficient is determined from the measured concentration data.

This procedure has been used since the first exploration of diffusion in the mid-19th-century:

- In 1850, Thomas Graham connected two bottles containing liquids with different concentrations to observe the transient diffusion process (cf. Figure 2.1(b)) (Graham, 1850; Cussler, 2007).
- In 1855: Adolf Fick used an open capillary with constant concentration gradient to verify his diffusion law (cf. Figure 2.1(c)) (Fick, 1855; Cussler, 2007).

Since then, many more sophisticated diffusion experiments have been developed, e.g.,

- the open-ended capillary method (cf. Figure 2.1(d)) (Anderson and Saddington,

1949),

- the Taylor dispersion method (Leaist, 1990; Price, 1988; van de Ven-Lucassen et al., 1997)
- the free diffusion method (Albright and Miller, 1975, 1980; Miller et al., 1996),
- microfluidic measurement methods (Kamholz et al., 1999; Häusler et al., 2012).

Diffusion experiments have even been conducted in space to avoid the influence of gravity (Shevtsova et al., 2014; Mialdun et al., 2013).

Even by use of the most advanced equipment, diffusion measurements are time-consuming and laborious: Typical diffusion measurements last from about an hour up to several days (Woolf et al., 1991). At the same time, diffusion measurements consume sample volumes ranging from a few mL to several 100 mL (Woolf et al., 1991; Gordon, 1945).

In addition, it is usually not sufficient to execute a single diffusion measurement. Instead, multiple diffusion measurements are required due to the following reasons:

1. Diffusion depends on temperature, pressure, and composition. The temperature and pressure levels are usually fixed during one experiment. If diffusion at multiple temperature and pressure levels is to be explored, multiple diffusion measurements are to be executed. The composition dependence is usually retrieved from multiple measurements with various initial mean concentrations $\bar{\boldsymbol{c}} = (\boldsymbol{c}_0 + \boldsymbol{c}_1)/2$ in the two diffusion domains (cf. Figure 2.1(a)).
2. Even for a single data point, diffusion measurements with multicomponent systems usually require multiple measurements: A multicomponent system containing N_c components requires the execution of $(N_\mathrm{c} - 1)$ diffusion measurements (Peters et al., 2016).
3. If statistical uncertainties should be evaluated and/or reduced, repetitions of the diffusion measurements are required.

Thus, the measurement of diffusion is burdened with high effort. Therefore, diffusion data are still scarce. In particular, there is a persisting demand for multicomponent diffusion coefficients (Economou et al., 2014; Shapiro, 2003; Bird, 2004): To date, diffusion coefficients of less than 300 multicomponent liquid systems are published. Less than 30 % of these systems are non-aqueous (Mutoru and Firoozabadi, 2011). For comparison: the commercial Dortmund Data Bank (DortmundDataBank, 2019) contains vapor-liquid equilibrium data on more than 2000 multicomponent mixtures. Here, more than 60 % of the multicomponent mixtures are non-aqueous.

To reduce the experimental effort, 2 main strategies are pursued:

1. Miniaturization: The diffusion time is proportional to the squared diffusion distance (Whitesides, 2006; Estévez-Torres et al., 2008). Therefore, the reduction of the geometrical dimensions results in reduced measurement times. In addition, small geometrical dimensions reduce the amount of required sample volume. An exemplary application of a miniaturized diffusion experiment is the use of microfluidic measurement methods which allow for measurement times of $< 1\,\text{h}$ and sample volumes of $\approx 10\,\mu\text{L}$ (e.g., Kamholz et al. (2001); Bouchaudy et al. (2018)).

2. Reduction of the number of required diffusion measurements. For this purpose, 3 methods are applied:

    a) Use of analytical techniques that allow for the simultaneous quantification of all components of a multicomponent mixture, e.g. Raman spectroscopy. Thereby, it is possible to determine multicomponent diffusion coefficients from a single experiment (e.g., Bardow et al. (2006)).

    b) Use of optimal experimental design (OED) techniques, which identify diffusion experiments with the highest information content, i.e. experiments leading to more accurate diffusion coefficients. Thereby, less diffusion experiments are required to obtain equally accurate diffusion coefficients (e.g., Häusler et al. (2012); Bardow et al. (2003)).

    c) Direct determination of concentration-dependent diffusion coefficients from a single experimental run. For this purpose, pure components are filled into the initial diffusion domains (cf. Figure 2.1(a)). During the subsequent diffusion process, the compositions run through the whole composition range. Thus, in principle, the full composition dependence of the diffusion coefficient can be estimated (Gupta and Cooper, 1971); fewer experiments are necessary, less sample is consumed, and sample preparation is much easier since pure components can be used instead of multiple pre-mixed mixtures which are difficult to prepare (e.g., Bardow et al. (2005a)).

Several combinations of these strategies have been implemented to reduce the experimental effort:

1. For binary liquid systems:
   - Concentration-dependent binary diffusion coefficients from a single experiment have been obtained from measurements in diaphragm cells (Clunie et al., 1990), ordinary diffusion cells (Durou et al., 1974; Bardow et al., 2005a; Kriesten et al., 2009), and microfluidic devices (Bouchaudy et al., 2018).
   - Microfluidics and Raman spectroscopy have been combined to obtain concentration-dependent binary diffusion coefficients from a single experiment with minimal sample consumption (Salmon et al., 2005; Bouchaudy et al., 2018).
   - Raman spectroscopy and OED techniques have been combined to obtain accurate binary diffusion coefficients from a minimum number of experiments in ordinary diffusion cells (Bardow et al., 2003).
   - Microfluidics (Werts et al., 2012; Kamholz et al., 2001; Hatch et al., 2001; Pappaert et al., 2005; Costin et al., 2013) and diffusion capillaries (Anderson and Saddington, 1949; Bacon and Adams, 1970) have been used to obtain binary liquid diffusion coefficients with minimum sample consumption.
   - Microfluidics and OED techniques have been combined to obtain accurate diffusion coefficients with minimum experimental effort and sample consumtion Häusler et al. (2012) combined UV-VIS spectroscopy and microfluidics with OED techniques to obtain accurate diffusion coefficients with minimum experimental effort.
2. For ternary liquid systems:
   - Raman spectroscopy has been combined with OED techniques to obtain ternary liquid diffusion coefficients from a single experiment (Bardow et al., 2006)
3. For binary gaseous systems:
   - Attempts have been made to also obtain concentration-dependent diffusion coefficients of gases from single experiments performed in a Loschmidt cell combined with holographic interferometry Kugler et al. (2013, 2015b);

Kullnick (2001); Baranski (2002); Buttig et al. (2011); Kugler et al. (2015a). However, in all these attempts, the identified concentration dependencies showed systematic errors and were in disagreement with literature values. Although a systematic study of Kugler et al. (2015a) could exclude many potential error sources, the reason for the observed systematic errors could not be fully resolved.

Thus, a number of successful implementations has been made to reduce the experimental effort. However, the implementations are still limited. In particular, there is a persisting demand for improved diffusion experiments for multicomponent systems, broader applications of OED techniques to other diffusion experiments than ordinary diffusion cells, and improved diffusion experiments for gases.

2.1.2 Modeling diffusion

A model-based analysis of diffusion experiments requires a mathematical description of the experiment. In this section, a brief overview on the modeling of diffusion experiments is given. For excellent detailed accounts on diffusion theory, the reader is referred to Taylor and Krishna (1993); Bird et al. (2002).

Physically, diffusion is the relative movement of molecules in a mixture. The molar flux $\boldsymbol{N}_i$ of species i can be described in terms its average velocity $\boldsymbol{u}_i$ with respect to a stationary coordinate reference frame as

$$\boldsymbol{N}_i = c_i \boldsymbol{u}_i, \quad i = 1, ..., N_\mathrm{c} \tag{2.1}$$

where c_i is the molar concentration of species i, and N_c is the total number of components. Commonly, the molar flux $\boldsymbol{N}_i$ is split into a *convective* and a *diffusive* flux:

$$\boldsymbol{N}_i = c_i \boldsymbol{u}^a + \boldsymbol{J}_i^a, \quad i = 1, ..., N_\mathrm{c}. \tag{2.2}$$

The convective flux $c_i \boldsymbol{u}^a$ describes the movement of the mixture as a whole, where each species moves with the same reference velocity $\boldsymbol{u}^a$. The choice of the reference frame a is arbitrary in principle and up to the modeler. However, clever, case-specific choices of reference frames exist (see below). The reference velocity $\boldsymbol{u}^a$ is the weighted average of the species' velocities $\boldsymbol{u}_i$:

$$\boldsymbol{u}^a = \sum_{i=1}^{N_c} a_i \boldsymbol{u}_i, \quad \sum_{i=1}^{N_c} a_i = 1 \tag{2.3}$$

where a_i are weighting factors. Important reference velocities (and their weighting factors) are the molar reference velocity u^{M} (with $a_i = x_i$), and the volume average velocity u^{V} (with $a_i = c_i \bar{V}_i$). Here, x_i is the mole fraction, and $\bar{V}_i$ is the partial molar volume of species i.

The diffusive flux $\boldsymbol{J}_i^a$ describes the relative movement of species i to the reference velocity $\boldsymbol{u}^a$. Therefore, the diffusive flux depends on the chosen reference frame. Depending on the chosen reference frame, complicated diffusion models can result which may include moving boundary problems (Bardow et al., 2003) and model indices > 1 (Martinson and Barton, 2000). Such complicated models are typically not suited for standard simulation software. In practice, it is advantageous to choose a reference frame a such that the convective reference velocity vanishes: $\boldsymbol{u}^a = \boldsymbol{0}$. Thereby, less complicated diffusion models are obtained that are more likely to be manageable by standard simulation software.

Of all N_c diffusive fluxes, only $N_c - 1$ diffusive fluxes are independent: Summing Equations 2.1 and 2.2 over all N_c components leads to the constraint

$$\sum_{i=1}^{N_c} \frac{a_i}{x_i} \boldsymbol{J}_i^a = 0. \tag{2.4}$$

In order to model the diffusive fluxes $\boldsymbol{J}_i^a$, two approaches are commonly used: The phenomenological law of Fick and the physically motivated Maxwell-Stefan (MS) approach. Both approaches are described and compared to each other in the following paragraphs.

Fick's law

Fick's law is based on the phenomenological observation that the diffusive flux in the volume reference frame $\boldsymbol{J}_i^{\mathrm{V}}$ is proportional to the spatial concentration gradient ∇c_j:

$$\boldsymbol{J}_i^{\mathrm{V}} = -\sum_{j=1}^{N_{\mathrm{c}}-1} D_{ij}^{\mathrm{V}} \boldsymbol{\nabla} c_j, \quad i = 1, ..., N_{\mathrm{c}} - 1. \tag{2.5}$$

Here, D_{ij}^{V} are the Fick diffusion coefficients in the volume reference frame.

Fick's law can also be formulated in the molar reference frame, where the spatial mole fraction gradient $\boldsymbol{\nabla} x_j$ is used as the driving force:

$$\boldsymbol{J}_i^{\mathrm{M}} = -c_{\mathrm{t}} \sum_{j=1}^{N_{\mathrm{c}}-1} D_{ij}^{\mathrm{M}} \boldsymbol{\nabla} x_j, \quad i = 1, ..., n_{\mathrm{c}} - 1. \tag{2.6}$$

Here, D_{ij}^{M} are the Fick diffusion coefficients in the molar reference frame.

The diagonal elements D_{ii}^a of the Fick diffusion coefficient matrix in reference frame a are the main diffusion coefficients describing the influence of the spatial gradient $\boldsymbol{\nabla} c_i$ or $\boldsymbol{\nabla} x_i$, respectively, of species i on its own diffusive flux $\boldsymbol{J}_i^a$. The off-diagonal elements D_{ij}^a are the cross diffusion coefficients describing the influence of the spatial gradient $\boldsymbol{\nabla} c_j$ or $\boldsymbol{\nabla} x_j$, respectively, of species j on the diffusive flux of component i. The off-diagonal elements D_{ij}^a can also be negative. The numerical values of D_{ij}^a depend on the numbering of the components of the system. Therefore, single values of D_{ij}^a cannot give unambiguous insight into the interactions of species i and j. Usually, the main contribution on the diffusive flux is provided by the diagonal elements, since $D_{ii}^a > D_{ij}^a$. The entire, usually non-symmetric matrix $\boldsymbol{D}^a$ is always positive definite.

The values of the $(N_{\mathrm{c}} - 1)^2$ Fick diffusion coefficients depend on the numbering of the components and on the chosen reference frame. Equations for the transformation of Fick diffusion coefficients from one reference frame to another are summarized in Taylor and Krishna (1993). For binary mixtures, the Fick diffusion coefficients in the volume reference frame and in the molar reference frame are identical: $D_{12}^{\mathrm{V}} = D_{12}^{\mathrm{M}}$.

Maxwell-Stefan equations

The Maxwell-Stefan (MS) equations are physically motivated and can be derived from the thermodynamics of irreversible processes. In the MS equations, the driving force for diffusion is the spatial gradient of the chemical potential, $\boldsymbol{\nabla} \mu_i$. The driving forces are balanced by friction forces which are induced by the differences $(\boldsymbol{u}_i - \boldsymbol{u}_j)$ between the average velocities of species i and j. The MS equations read as

$$-\frac{1}{RT}\boldsymbol{\nabla}\mu_i|_{T,p} = \sum_{j=1,j\neq i}^{N_c-1} \frac{x_i x_j (\boldsymbol{u}_i - \boldsymbol{u}_j)}{Đ_{ij}}, \quad i = 1, ..., N_c - 1.. \tag{2.7}$$

Here, R is the gas constant, T is the absolute temperature, and μ_i is the chemical potential of species i at constant temperature T and pressure p. $Đ_{ij}$ are the MS diffusion coefficients which can be interpreted as inverse friction coefficients. The MS diffusion coefficients are independent of the chosen reference frame and are always symmetric: $Đ_{ij} = Đ_{ji}$.

In practical applications, chemical potentials are usually not directly accessible. Therefore, it is convenient to reformulate the MS equations into a form analogous to Fick's law in the molar reference frame:

$$\boldsymbol{J}^{\mathrm{M}} = -c_t \boldsymbol{B}^{-1}\boldsymbol{\Gamma}\boldsymbol{\nabla}\boldsymbol{x}. \tag{2.8}$$

Here, $\boldsymbol{J}^{\mathrm{M}}$ and $\boldsymbol{x}$ are the conjunctions of the diffusive fluxes $\boldsymbol{J}_i^{\mathrm{M}}$ and the mole fractions x_i, respectively. The elements of the matrix $\boldsymbol{B}$ are given by

$$B_{ii} = \frac{x_i}{Đ_{in_c}} + \sum_{j=1,j\neq i}^{N_c} \frac{x_j}{Đ_{ij}} \tag{2.9}$$

and

$$B_{ij} = -x_i \left(\frac{1}{Đ_{ij}} - \frac{1}{Đ_{iN_c}}\right). \tag{2.10}$$

The matrix $\boldsymbol{\Gamma}$ is the so-called *thermodynamic factor*. It contains thermodynamic information of the system and its elements are defined by

$$\Gamma_{ij} = \delta_{ij} + x_i \left(\frac{\partial \ln \gamma_i}{\partial x_j}\right)_{T,p,\Sigma}. \tag{2.11}$$

Here, δ_{ij} denotes the Kronecker delta and γ_i denotes the activity coefficient of species i. The differentiation of the activity coefficient is carried out at constant temperature T and pressure p, and at constant mole fraction of all other components except the N_cth one, so that $\sum_i^{N_c} x_i = 1$ during the differentiation (indicated by the symbol Σ).

The thermodynamic factor is a measure for the nonideality of a mixture (Taylor and Krishna, 1993). For ideal mixtures and pure substances, the thermodynamic factor is equal to the identity matrix: $\mathbf{\Gamma} = I$. Usually, the thermodynamic factor is not symmetric.

Comparing Fick's law and the Maxwell-Stefan equations

Since both Fick's and the MS approach describe the same phenomenon, they are linked to each other. By comparing Equations 2.6 and 2.8, an expression for the relation between MS and Fick diffusion coefficients can be derived:

$$\boldsymbol{D}^{\mathrm{M}} = \boldsymbol{B}^{-1}\mathbf{\Gamma}. \tag{2.12}$$

For an ideal mixture, $\mathbf{\Gamma} = \boldsymbol{I}$ and the Fick and MS diffusion coefficient matrices coincide: $\boldsymbol{D}^{\mathrm{M}} = \boldsymbol{B}^{-1}$. In the binary case, Equation 2.12 reduces to the scalar equation

$$D_{12} = Đ_{12}\Gamma. \tag{2.13}$$

A comparison between the approaches of Fick and MS shows the following important characteristics:

- The MS approach separates thermodynamic effects ($\mathbf{\Gamma}$) from binary molecular interactions ($\boldsymbol{B}$), whereas Fick's approach lumps both effects into the Fick diffusion coefficient $\boldsymbol{D}^{\mathrm{M}}$ (cf. Equation 2.12).
- While Fick diffusion coefficients depend on the velocity reference frame and the component numbering, MS diffusivities do not.
- While the MS diffusivity matrix is symmetric, the Fick diffusivity matrix is not. Hence, $N_{\mathrm{c}}\,(N_{\mathrm{c}} - 1)$ MS diffusivities are sufficient whereas $(N_{\mathrm{c}} - 1)^2$ Fick diffusivities are needed to describe an N_{c}-component mixture. Therefore, for $N_{\mathrm{c}} > 2$, the Fick diffusivities are not independent.
- Both MS and Fick diffusion coefficients depend on temperature, pressure, and composition. However, MS diffusivities are typically found to be less concentration-dependent than Fick diffusivities (Taylor and Krishna, 1993). The increased concentration dependence of the Fick diffusivities results from the concentration dependence of the thermodynamic factor (cf. Equation 2.12).

In principle, both approaches of Fick and MS are on a par. However, the comparison shows that the MS approach provides a thermodynamically more sound basis for the calculation of diffusion. Therefore, the MS approach is particularly suited (and typically used) to *predict* diffusion coefficients (Taylor and Krishna, 1993). For this purpose, models for the MS diffusion coefficients and the thermodynamic factors are combined to obtain Fick diffusion coefficients. However, the application of the MS approach is often limited in practice: On the one hand, the direct measurement of the chemical potential gradient is practically impossible; on the other hand, the calculation of the thermodynamic factor still introduces large uncertainties even with today's Gibbs free energy (g^{E})-models and equations of state. This dilemma exists since the derivation of the MS equations 150 years ago. In contrast, the phenomenological approach of Fick is better suited to *measure* diffusion coefficients since only concentrations and no chemical potentials are directly measurable. In Section 2.1.3, the prediction of diffusion coefficients is discussed in more detail.

Diffusion process models

To model a diffusion experiment, the molar fluxes $\boldsymbol{n}_i$ (cf. Equation 2.1) have to be embedded into an overall mass balance. The component mole balance reads as

$$\frac{\partial c_i}{\partial t} + \boldsymbol{\nabla} \boldsymbol{n}_i = \boldsymbol{0} \tag{2.14}$$

where t is the time. Insertion of Equation 2.1 leads to

$$\frac{\partial c_i}{\partial t} + \boldsymbol{\nabla}(c_i \boldsymbol{u}^a + \boldsymbol{J}_i^a) = \boldsymbol{0} \tag{2.15}$$

The complexity of Equation 2.15 (and thereby the diffusion process model) depends strongly on the chosen reference frame a.

For **ideal systems**, the volume of the mixture is constant during the diffusion process. Thus, excess volume and the volume reference velocity are zero: $V^{\mathrm{E}} = 0$, $\boldsymbol{u}^{\mathrm{V}} = 0$. Therefore, the volume reference frame $a = \mathrm{V}$ is a clever choice. Insertion of Fick's law in the volume reference frame (Equation 2.5) into Equation 2.15 leads to

$$\frac{\partial c_i}{\partial t} = \boldsymbol{\nabla} \sum_{j=1}^{N_{\mathrm{c}}-1} D_{ij}^{\mathrm{V}} \boldsymbol{\nabla} c_j. \tag{2.16}$$

Alternatively, the MS Equation 2.7 could have been inserted into Equation 2.15.

If a constant diffusion coefficient is assumed, Equation 2.16 further simplifies to

$$\frac{\partial c_i}{\partial t} = \sum_{j=1}^{N_c - 1} D_{ij}^{V} \boldsymbol{\nabla}^2 c_j \tag{2.17}$$

which is also known as *Fick's second law.*

For **real systems**, the volume of the mixture changes during the diffusion process. Thus, the excess volume and the volume reference velocity are non-zero: $\boldsymbol{u}^{V} \neq \boldsymbol{0}$, $\boldsymbol{u}^{V} \neq 0$. However, the total number of moles remains constant. Therefore, it is more convenient to use the molar reference frame $a = \mathrm{M}$.

Consider a 1-dimensional diffusion process. The component mole balance in laboratory coordinates z then reads as

$$\frac{\partial c_i(z,t)}{\partial t} + \frac{\partial n_i(z,t)}{\partial z} = 0 \tag{2.18}$$

Laboratory coordinates imply differential elements with equal volumes $\mathrm{d}V$. Thus, laboratory coordinates are linked to the volume reference frame. To transform the total mole balance from the volume reference frame into the molar reference frame, we have to use differential elements with equal numbers of moles $\mathrm{d}n$. $\mathrm{d}n$ and $\mathrm{d}V$ are related by

$$\mathrm{d}n = c_t(z,t)\, A\, \mathrm{d}z. \tag{2.19}$$

For 1-dimensional diffusion, the differential elements may be interpreted as "slices" perpendicular to the direction of diffusion. In the volume reference frame, the slices have equal heights $\mathrm{d}z = \mathrm{d}V/A$, where A is the cross-sectional area of the diffusion cell. In the molar reference frame, the slices have equal numbers of moles $\mathrm{d}n$. For convenience, $\mathrm{d}n$ can be transformed into a new coordinate $\mathrm{d}\bar{z}$ with units of length by use of an arbitrary molar volume, e.g. the molar volume V_2^0 of reference species 2 (Crank, 1975):

$$\mathrm{d}\bar{z} = c_t(z,t)\, V_2^0\, \mathrm{d}z \tag{2.20}$$

and after integration

$$\bar{z} = \int_0^z c_t(z,t)\, V_2^0\, \mathrm{d}z'. \tag{2.21}$$

The molar concentration in the transformed system, ξ_i, is

$$\xi_i^{\mathrm{M}}(\bar{z},t) = \frac{\mathrm{d}n_i}{\mathrm{d}\bar{V}} = \frac{\mathrm{d}n_i}{A\,\mathrm{d}\bar{z}} = \frac{1}{V_2^0}\frac{\mathrm{d}n_i}{dn} = \frac{x_i(z,t)}{V_2^0}. \tag{2.22}$$

Thus, Eqs. 2.21 and 2.22 allow for the transformation from the volume reference frame to the molar reference frame and vice versa. Combining Eqs. 2.18, 2.21, 2.22, and introducing Fick's law in the molar reference frame (Equation 2.6) leads to the differential equation

$$\frac{\partial \xi_i^{\mathrm{M}}(\bar{z},t)}{\partial t} = \frac{\partial}{\partial \bar{z}}\left(\sum_{j=1}^{n_c-1} D_{ij}^{\mathrm{M}}\,(c_t(\bar{z},t)\,V_2^0)^2\,\frac{\partial \xi_i^{\mathrm{M}}(\bar{z},t)}{\partial \bar{z}}\right). \tag{2.23}$$

A more detailed derivation of the transformation from the volume reference frame to the molar reference frame is provided in literature (Bardow et al., 2005b; Crank, 1975; Bausa and Marquardt, 2001). Eq. 2.23 does not contain a convective velocity but is a simple diffusion equation which can be solved easily with standard simulation software.

To complete the diffusion model (Equation 2.16 or 2.23), initial and boundary conditions as well as constitutive equations have to be introduced. These equations depend on the specific diffusion process. Therefore, these case-specific equations will be introduced in the corresponding chapters.

2.1.3 Predicting diffusion coefficients

The prediction of Fick diffusion coefficients is usually based on the thermodynamically sound MS approach (cf. Section 2.1.2). Fick diffusion coefficients are then obtained using the following procedure:

1. prediction of MS diffusion coefficients $Đ_{ij}$ and subsequent calculation of the matrix $\boldsymbol{B}$ (Equations 2.9 and 2.10),
2. prediction of the matrix of thermodynamic factors, $\boldsymbol{\Gamma}$ (Equation 2.11),
3. calculation of the Fick diffusion coefficient matrix $\boldsymbol{D}$ (Equation 2.12).

In principle, two methods are available to obtain MS diffusion coefficients: Predictions based on Molecular Dynamics (MD) simulations, and predictions based on engineering models. Both methods are described in the following paragraphs. Subsequently, we discuss the prediction of thermodynamic factors from Molecular Dynamics simulations, equations of state, and excess Gibbs energy models.

Predicting MS diffusion coefficients from Molecular Dynamics simulations

Due to increasing computational power, the use of Molecular Dynamics (MD) simulations to predict MS diffusion coefficients has gained increasing attention in the last decade. MD simulations have been performed to study diffusion in both gases (Makrodimitri et al., 2015, 2011; Moultos et al., 2014, 2016; Heller et al., 2016, 2014; Koller et al., 2015) and liquids (Liu et al., 2011c,b, 2012, 2011a; Guevara-Carrion et al., 2016b, 2008; Parez et al., 2013). There are many excellent textbooks on MD simulations techniques (Frenkel and Smit, 2002; Allen and Tildesley, 2017; Tuckerman, 2010; Rapaport, 2004). In the following, we briefly outline the idea and principle of MD simulations for the calculation of MS diffusion coefficients.

The central idea of MD simulations is that atoms/molecules interact with each other; these interactions either follow from quantum mechanics or a parameterized functional form (the so-called force field).The forces on atoms/molecules that follow from these interactions are used to integrate Newton's second law (force equals mass times acceleration) numerically. In this way, trajectories are obtained for all the atoms and molecules in the system. Essentially, the result of an MD simulation is a "movie" that shows how all atoms and molecules move around as a function of time. From such a "movie", average properties of the system can be computed: (1) static (thermodynamic) properties, e.g., the thermodynamic factor $\mathbf{\Gamma}$; and (2) dynamic properties (properties related to the time evolution of the system), e.g., MS diffusion coefficients $Đ_{ij}$.

MS diffusion coefficients are often calculated from *equilibrium* MD simulations which employ three-dimensional isotropic systems at equilibrium. Expressions for the MS diffusion coefficient can be obtained by applying the linear-response theory (Evans and Morriss, 2008) where the MS diffusion coefficients are calculated from time-correlation functions of the molecular trajectories. Three types of correlation functions can be distinguished:

1. Self-correlation functions: The self-correlation functions are a measure for the

mean-square displacements of individual atoms/molecules. The self-correlation functions correspond to the self-diffusion coefficients $D_{i,\text{self}}$ (also called tracer diffusion coefficients or intradiffusion coefficients in literature):

$$D_{i,\text{self}} = \lim_{t\to\infty} \frac{1}{6N_i t} \left\langle \sum_{j=1}^{N_i} \left(\boldsymbol{r}_{j,i}(t) - \boldsymbol{r}_{j,i}(0)\right)^2 \right\rangle . \tag{2.24}$$

Here, t is the correlation time, N_i is the number of atoms/molecules of species i and $\boldsymbol{r}_{j,i}$ is the position of the j-th molecule of species i. The angle brackets denote the ensemble average.

2. Cross-correlation functions between atoms/molecules of the same species i:

$$CC_{ii} = \lim_{t\to\infty} \frac{1}{6N_\text{t} t} \left\langle \left(\sum_{k=1}^{N_i} \left(\boldsymbol{r}_{k,i}(t) - \boldsymbol{r}_{k,i}(0)\right) \right) \cdot \left(\sum_{l=1,l\neq k}^{N_i} \left(\boldsymbol{r}_{l,i}(t) - \boldsymbol{r}_{l,i}(0)\right) \right) \right\rangle . \tag{2.25}$$

Here, N_t is the total number of atoms/molecules in the simulation box.

3. Cross-correlation functions between atoms/molecules of different species i and j:

$$CC_{ij} = \lim_{t\to\infty} \frac{1}{6N_\text{t} t} \left\langle \left(\sum_{k=1}^{N_i} \left(\boldsymbol{r}_{k,i}(t) - \boldsymbol{r}_{k,i}(0)\right) \right) \cdot \left(\sum_{l=1}^{N_j} \left(\boldsymbol{r}_{l,j}(t) - \boldsymbol{r}_{l,j}(0)\right) \right) \right\rangle . \tag{2.26}$$

All three types of correlation functions jointly determine the MS diffusion coefficient. For a binary mixture, the MS diffusion coefficient $Đ_{12}$ is

$$Đ_{12} = \underbrace{x_2 D_{1,\text{self}} + x_1 D_{2,\text{self}}}_{Đ_\text{Darken}} + \underbrace{x_1 x_2 N(CC_{11} + CC_{22} - 2CC_{12})}_{Đ_\text{Cross}} . \tag{2.27}$$

The MS diffusion coefficient $Đ_{12}$ is composed of two parts: An (ideal) Darken diffusion coefficient $Đ_\text{Darken}$ containing the self-diffusion coefficients $D_{i,\text{self}}$, and a non-ideal diffusion coefficient $Đ_\text{Cross}$ containing the cross-correlation functions CC_{ij}. For approximately ideal mixtures with weak molecular interactions, the atoms/molecules move randomly. Thus, the cross-correlations CC_{ii} and CC_{ij} are negligible compared to the self-diffusion coefficients $D_{i,\text{self}}$ and the MS diffusion coefficient is approximately the Darken diffusion coefficient: $Đ_{12} \approx Đ_{\text{Darken}}$. For non-ideal mixtures with strong molecular interactions, the atoms/molecules do not move randomly but can form groups or clusters. Thus, the cross-correlations CC_{ii} and CC_{ij} and the non-ideal diffusion coefficient $Đ_{\text{Cross}}$ can be in the same order of magnitude as the Darken diffusion coefficient $Đ_{\text{Darken}}$ ((Liu et al., 2011d)): For self-associating mixtures, the velocity cross-correlations CC_{ii} between atoms/molecules of the same component i become predominant; for mixtures with associating behavior between unlike atoms/molecules of different components i and j, the velocity crosscorrelations CC_{ij} become predominant. For multicomponent mixtures, qualitatively similar, but more complicated expressions apply (see Liu et al. (2011d)).

The values of the transport coefficients depend on the box size of the MD simulations. Therefore, it is important to correct MD simulation results for these so-called *finite-size effects*. Details on correction strategies have been developed by the present author and colleagues and can be found, e.g., in Jamali et al. (2018).

A key advantage of MD simulations is that we can get detailed insight into the effect of molecular interactions on the properties of a system: Equation 2.27 shows that MD simulations allow not only for the calculation of MS diffusion coefficients, but for a detailed analysis of the different contributions of the correlation functions on the MS diffusion coefficient. In addition, MD simulations provide not only diffusion coefficients, but a full set of transport and thermodynamic data: MD simulations provide both static properties like, e.g., the average total energy, and the average total pressure, as well as dynamic properties like, e.g., diffusivity, heat conductivity, and viscosity. Furthermore, MD simulations are very versatile: Different molecular interactions as well as different temperature and pressure conditions can be taken into account easily.

However, MD simulations always require reliable force fields which are not always available. Furthermore, MD simulations are often computationally too expensive to be performed in the framework of process simulations, where iterative solution schemes are applied. Therefore, predictive models have been developed which allow for a rapid estimation of diffusion coefficients. In the next paragraph, we give a short overview on state-of-the-art predictive models.

Predicting MS diffusion coefficients by engineering models

Engineering models aim at a rapid estimation of diffusion coefficients based on a minimum amount of data, e.g., diffusion coefficients at infinite dilution (Poling et al., 2001). The diffusion coefficients at infinite dilution $Đ_{ij}^{x_k \to 1}$ can be obtained from MD simulations (Liu et al., 2011c), diffusion experiments (Peters et al., 2016), or further predictive models (Poling et al., 2001).

In general, diffusion coefficients depend on temperature, pressure, and composition:

- Gaseous diffusion coefficients depend strongly on temperature and pressure, but less on composition (Poling et al., 2001; Marrero and Mason, 1972). Gaseous diffusion coefficients can be predicted with high accuracy from the Chapman-Eskog theory (Marrero and Mason, 1972; Poling et al., 2001; Jäger and Bich, 2017; Sharipov and Benites, 2017). If simple engineering equations are needed, correlations for specific mixtures are widely availabe (Winkelmann, 2007; Marrero and Mason, 1972).
- Liquid diffusion coefficients depend strongly on temperature and composition, but almost not on pressure (Poling et al., 2001; Marrero and Mason, 1972). Due to stronger, specific molecular interaction, theroretical predictions of liquid diffusion coefficients are much less accurate than predictions of gaseous diffusion coefficients (Poling et al., 2001).

The **temperature dependence** of gaseous diffusion coefficients is approximately given by (Poling et al., 2001)

$$Đ_{ij} \propto T^{1.5} \text{ to } T^2. \tag{2.28}$$

The exponent varies with temperature itself and has to be determined for specific temperature ranges.

The temperature dependence of liquid diffusion coefficients is roughly given by

$$Đ_{ij} \propto A \cdot \exp -\frac{B}{T}, \tag{2.29}$$

where A and B are empirical parameters which are highly case-specific (Poling et al., 2001).

The **pressure dependence** of gaseous diffusion coefficients is well-approximated by (Cussler, 2007)

$$Đ_{ij} \propto 1/p. \tag{2.30}$$

Only high-pressure applications show different pressure dependencies (Poling et al., 2001).

The **concentration dependence** of diffusion coefficients is very complicated and has attracted great attention in the last decades (Poling et al., 2001; Rehfeldt and Stichlmair, 2007; Peters et al., 2016; Liu et al., 2011c,d; Kooijman and Taylor, 1991). Within the numerous proposed engineering models, two main classes can be identified:

The first class of models is based on the Vignes equation (Vignes, 1966)

$$Đ_{12} = \left(Đ_{12}^{x_1 \to 1}\right)^{x_1} \left(Đ_{12}^{x_2 \to 1}\right)^{x_2}. \tag{2.31}$$

Here, $Đ_{12}$ is the (mutual) Maxwell-Stefan (MS) diffusion coefficient, x_1 and x_2 are the mole fractions of components 1 and 2, and $Đ_{12}^{x_1 \to 1}$ and $Đ_{12}^{x_2 \to 1}$ are the MS diffusion coefficients at infinite dilution. The Vignes equation is very popular since it only requires diffusion coefficients at infinite dilution as input for which many predictive models are available such as e.g. the Wilke-Change equation (Poling et al., 2001). The Vignes equation is purely empirical and applicable to binary systems only. However, extensions of the Vignes equation to multicomponent mixtures have been proposed (Kooijman and Taylor, 1991; Liu et al., 2011d).

The second class of models is based on the Darken equation (Darken, 1948)

$$Đ_{12} = Đ_{\text{Darken}} = x_2 D_{1,\text{self}} + x_1 D_{2,\text{self}}. \tag{2.32}$$

Here, $D_{1,\text{self}}$ and $D_{2,\text{self}}$ are the composition-dependent self-diffusion coefficients of components 1 and 2 in the mixture. The Darken equation has been extended to multicomponent mixtures by Liu et al. (Liu et al., 2011d,b). In contrast to the Vignes equation (Equation 2.31), the Darken equation has a physical basis and can be derived from statistical-mechanical theory when cross-correlations between the molecule trajectories are neglected, i.e. when the molecules in the mixture move independently and not in groups or clusters (cf. Equation 2.27 and the following paragraph). Hence, the Darken equation is suitable for ideal mixtures, but not for strongly non-ideal mixtures.

A modified Darken equation for non-ideal binary mixtures has therefore been proposed by D'Agostino et al. (2011) and Moggridge (2012b) which is based on critical point scaling laws (De et al., 2001; Clark and Rowley, 1988; Wu et al., 1988; Matos Lopes et al., 1992):

$$Đ_{12} = Đ_{\text{Darken}} \Gamma^{-0.36}. \tag{2.33}$$

Here, the Darken equation is corrected by a power function of the thermodynamic factor Γ, which is a measure for the nonideality of the system (cf. Equation 2.11). In the rest of this thesis, Equation 2.33 is called the "Moggridge equation". The Moggridge equation has been tested and successfully validated for a wide range of non-ideal liquid mixtures; average deviations between 12 %-15 % were observed (D'Agostino et al., 2011; Moggridge, 2012b; D'Agostino et al., 2012). The Moggridge equation is not applicable to mixtures with dimerizing components. Moggridge (2012a) proposed a further modification of Equation 2.33 for mixtures with dimerizing components. Zhu et al. (2015) introduced local mole fractions into the Moggridge equation. Thereby, mixtures with and without dimerizing species could be successfully described. Recently, Allie-Ebrahim et al. (2018) suggested an extension of the Moggridge equation to multicomponent systems.

Despite its sound physical background, the Darken equation (and thereby the Moggridge equation) is generally seen as "of little practical use due to the fact that it relies on the self-diffusion coefficients [$D_{i,\text{self}}$] in the mixture, which are rarely available" (Guevara-Carrion et al., 2008) To avoid the use of $D_{i,\text{self}}$, modifications of the Darken equation have been proposed which use self-diffusion coefficients at infinite dilution $D_{i,\text{self}}^{x_{j\neq i}\to 1} = D_{ij}^{x_{j\neq i}\to 1} = Đ_{12}^{\infty}$ (Rathbun and Babb, 1966), include additional, system-dependent modification factors (Sanchez and Clifton, 1977), or incorporate the shear viscosity (Anderson and Babb, 1961; Bidlack and Anderson, 1964), to name a few. Similar modifications have been applied to the Vignes equation to extend the applicability to a wider range of non-ideal systems (Leffler and Cullinan, 1970). However, the applicabilities of the (modified) Vignes equations and of those modified Darken equations that avoid the use of $D_{i,\text{self}}$ are very case specific (Poling et al., 2001; Taylor and Krishna, 1993). It was concluded that "no single correlation [that avoids the use of composition-dependent self-diffusion coefficients $D_{i,\text{self}}$] is always satisfactory for estimating the concentration effect on liquid diffusion coefficients" (Poling et al., 2001). Therefore, the composition-dependent self-diffusion coefficients should preferentially not be replaced in the Darken-based models.

Thus, reliable predictions of composition-dependent *self-diffusion coefficients* $D_{i,\text{self}}$ are needed. Existing predictive models for $D_{i,\text{self}}$ predict $D_{i,\text{self}}$ from the self-diffusion coefficients at infinite dilution, $D_{i,\text{self}}^{x_j\to 1}$. The self-diffusion coefficients $D_{i,\text{self}}^{x_j\to 1}$ can be obtained, e.g., from NMR measurements, diffusion measurements with radioactive tracers, or Molecular Dynamics simulations. In addition, the self-diffusion coefficients $D_{i,\text{self}}^{x_{j\neq i}\to 1}$ can be estimated with predictive models (Poling et al., 2001; Suarez-Iglesias et al., 2007), or from the binary mutual diffusion coefficients at infinite dilution (Krishna, 2015): $D_{i,\text{self}}^{x_{j\neq i}\to 1} = D_{ij}^{x_{j\neq i}\to 1} = D_{12}^{\infty}$. Extensive datasets on experimental self-diffusion coefficients at infinite dilution, $D_{i,\text{self}}^{x_{j\neq i}\to 1}$, do already exist (Suarez-Iglesias et al., 2015). A further method to obtain self-diffusion coefficients at infinite dilution is entropy scaling (Hopp et al., 2018).

Carman and Stein (1956) proposed the semi-empirical relation

$$D_{i,\text{self,pred}} = \frac{D_{i,\text{self}}^{x_j\to 1}\,\eta^{x_j\to 1}}{\eta} \tag{2.34}$$

for binary systems, where $D_{i,\text{self,pred}}$ is the predicted value of $D_{i,\text{self}}$, $\eta^{x_j\to 1}$ is the viscosity of pure component j, and η is the viscosity of the mixture. Equation 2.34 is based on the Stokes-Einstein equation (Poling et al., 2001) and works well for ideal mixtures (Carman and Stein, 1956). For non-ideal mixtures, large deviations occur (Guevara-Carrion et al., 2016b).

For non-ideal mixtures, Krishna and van Baten (2005) suggest the empirical relation

$$D_{i,\text{self,pred}} = \sum_{j=1}^{n} w_j D_{i,\text{self}}^{x_j\to 1}, \tag{2.35}$$

where w_j is the mass fraction of component j. Equation 2.35 was successfully tested for linear alkanes (Krishna and van Baten, 2005) and mixtures with thermodynamic factors $0.55 \leq \Gamma \leq 1$ (Guevara-Carrion et al., 2016b). For strongly non-ideal mixtures with thermodynamic factors $\Gamma < 0.55$, large deviations were observed (Guevara-Carrion et al., 2016b).

Based on derivations of Curtiss and Hirschfelder (1949) and Hirschfelder and Curtiss

(1949), McCarty and Mason (1960) and Miller and Carman (1961) derived the relation

$$\frac{1}{D_{i,\text{self,pred}}} = \frac{x_1}{D_{i,\text{self}}^{x_1 \to 1}} + \frac{x_2}{D_{i,\text{self}}^{x_2 \to 1}}, \quad i = 1, 2. \tag{2.36}$$

for binary gas mixtures. In the rest of this thesis, Equation 2.36 will be called the "McCarty-Mason equation". The McCarty-Mason equation is based on the assumption of an approximately constant mutual diffusion coefficient, which is an often valid assumption for gases. It is exact in the limit of infinite dilution (Liu et al., 2011d). McCarty and Mason (1960) and Miller and Carman (1961) tested the McCarty-Mason equation successfully with data from gas diffusion experiments. Liu et al. (2011d) proposed the use of the McCarty-Mason equation for weakly non-ideal liquids. Satisfying predictions with average relative deviations of 10 % of self-diffusion coefficients in weakly non-ideal liquids were observed (Liu et al., 2011d,b; Guevara-Carrion et al., 2016b). However, Equation 2.36 performs poorly for strongly non-ideal mixtures; average relative deviations of 25 % were observed (Guevara-Carrion et al., 2016b).

Thus, there is a need for improved predictions of concentration-dependent self-diffusion coefficients of non-ideal liquids. However, accurate predictions of self-diffusion coefficients are only one prerequisite for the successful prediction of Fick diffusion coefficients. In addition, reliable thermodynamic factors are required (cf. Equations 2.13 and 2.33). In the next paragraph, the state of the art to predict thermodynamic factors is introduced.

Predicting thermodynamic factors

In principle, the thermodynamic factor $\mathbf{\Gamma}$ can be predicted using any approach providing derivatives of the activity coefficients with respect to composition, $(\partial \ln \gamma_i / \partial x_j)$ (cf. Equation 2.11). In practice, Gibbs free energy (g^{E}) models are typically used (Peters et al., 2016) (cf. Appendix E). In recent years, the calculation of thermodynamic factors from molecular simulations has become an attractive aternative method (Peters et al., 2016).

g^{E} models are mostly used for the calculation of thermodynamic factors $\mathbf{\Gamma}$ (Taylor and Krishna, 1993; Guevara-Carrion et al., 2016b,a; Kooijman and Taylor, 1991). The g^{E} model parameters are typically obtained from vapor-liquid equilibrium (VLE) data (Kooijman and Taylor, 1991). VLE require an accurate description of the absolute values of the activity coefficients γ_i. However, even if the absolute values of γ_i are well-described, the derivatives $\partial \ln \gamma_i / \partial x_j$ may be ill-determined. On the one hand, the accuracy of the VLE data has a major influence, on the other hand, the model structure of the specific g^{E} model itself determines the derivative $\partial \ln \gamma_i / \partial x_j$ - and thereby the thermodynamic factor $\mathbf{\Gamma}$. Composition dependencies of $\mathbf{\Gamma}$ that do not only differ in magnitude, but also in shape have been reported (Kooijman and Taylor, 1991; Guevara-Carrion et al., 2016b). Therefore, the careful selection and fit of experimental VLE data to g^{E} models has been a major task in estimating the thermodynamic factor (Guevara-Carrion et al., 2016b,a; Zhu et al., 2015; D'Agostino et al., 2013; Moggridge, 2012b).

In molecular simulations, the thermodynamic factor can be calculated from the permuted Widom test particle insertion method (Balaji et al., 2013) or Kirkwood-Buff integrals (Krüger et al., 2013). Both methods have been used successfully with average deviations of the thermodynamic factor of approximately 5 % (Parez et al., 2013; Liu et al., 2013, 2011b,a, 2012).

The uncertainty in the thermodynamic factor obtained from G^{E} models has hindered the establishment of the MS approach to predict Fick diffusion coefficients. However, the recent progress in the calculation of thermodynamic factors from MD simulations seems to be promising. Still, MD simulations are often computationally too expensive to be performed in the framework of process simulations and require reliable force fields which are not always available.

2.2 Liquid-liquid equilibria: Extraction

Liquid-liquid extraction is a well-established separation procedure in the laboratory as well as in food, pharmaceutical, and chemical industry. Liquid-liquid extraction is based on the occurrence of liquid-liquid equilibra (LLE). LLE result from significantly different molecular interactions between the different species of a mixture. As an example, consider a ternary system consisting of species 1, 2, and 3 (cf. Figure 2.2(a)). If the attractive interactions between different species $(1-3)$ are much weaker than the interactions between the same species $(1-1$ and $3-3)$, the mixture will separate into two liquid phases. One phase is mainly composed of species 1, while the other phase is mainly composed of species 3. Depending on the mass transport velocity, the mixture will reach an equilibrium state sooner or later. Now consider a species 2 that has strong attractive molecular interactions with species 1. If a mixture of species 3 and 2 is brought into contact with species 1, species 1 and 3 will separate and species 2 will be transferred from species 3 to species 1; in other words: species 2 is extracted from 3.

Due to the complex molecular interactions, the prediction of LLE is a major challenge and burdened with large uncertainty (Case et al., 2011). Therefore, experimental investigations are the most significant source for LLE data (Kontogeorgis, 2013; Hendriks et al., 2010). In the following Section 2.2.1, we introduce the basic principle of LLE measurements. Subsequently, we describe the mathematical description of LLE (Section 2.2.2).

2.2.1 Measuring liquid-liquid equilibria

LLE data are typically measured by equilibrating two liquid phases in an equilibrium cell. Figure 2.2(b) illustrates the basic construction of an LLE experiment: In a first step, a mixture with an overall composition of components, $\boldsymbol{z}$, is prepared. Subsequently, this mixture separates into two phases $'$ and $''$ with phase compositions $\boldsymbol{x}'$ and $\boldsymbol{x}''$. To determine these phase compositions $\boldsymbol{x}'$ and $\boldsymbol{x}''$, two methods are commonly employed:

1. Using an analytical method, samples are drawn from each phase and the sample compositions are analyzed using an appropriate analytical technique. Commonly chromatographic methods are applied (Bolden et al., 2002; Kuzmanović et al., 2003; Alimuddin et al., 2008; Dechambre et al., 2014a).
2. Using the turbidity (or clound point) method, the composition of the initial mixture has to be *known*. The initial mixture may be homogeneous or heterogeneous. Subsequently, a mixture of known composition is added to the initial mixture until phase separation appears or disappears. At this moment, the composition of the predominant phase is equal to the overall composition: $\boldsymbol{x}' = \boldsymbol{z}$. Since the overall composition $\boldsymbol{z}$ is known, the phase composition $\boldsymbol{x}'$ is known as well. The phase composition $\boldsymbol{x}''$ can be determined analogously

Typically, the equilibrium cells used for the measurement of LLE have relatively large volumes ($\approx 25\,\text{mL}$). Therefore, large sample volumes are needed. At the same time,

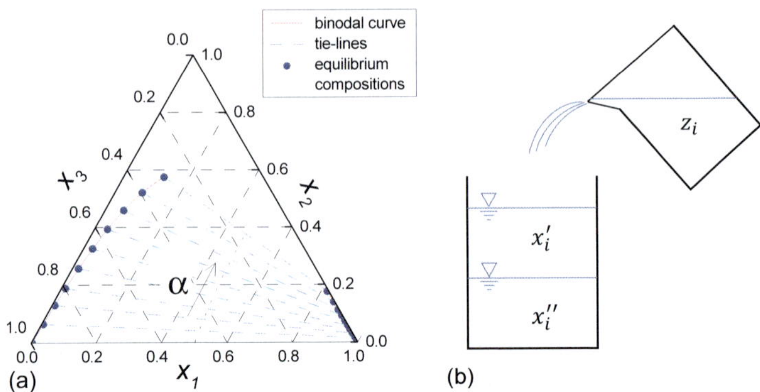

Figure 2.2: **(a)** Ternary phase diagram of a type I LLE.
(b) Schematic illustration of an LLE experiment.

since equilibration is based on diffusion, long equilibration times are needed (cf. Section 2.1.1). Thus, typical LLE measurements last from an hour up to several days (Dechambre et al., 2014a).

In addition, it is usually not sufficient to execute a single LLE measurement. Instead, multiple LLE measurements are required due to the following reasons:

1. LLE depend strongly on temperature. Thus, multiple LLE measurements in the neighborhood of the design temperature are typically executed.
2. For multi-stage extraction processes, partial or full measurement of the binodal curve is required. Thus, multiple LLE experiments have to be executed.
3. If statistical uncertainties should be evaluated and/or reduced, repetitions of the LLE measurements are required.

Thus, the measurement of full LLE is burdened with large sample consumption and long equilibration times. To overcome these limitations, there have been attempts to automate and miniaturize experiments (Carlsson and Karlberg, 2000; Bolden et al., 2002; Kuzmanović et al., 2003; Alimuddin et al., 2008; Marine et al., 2009; Dechambre et al., 2014a; Thien et al., 2017, 2019). However, with shorter equilibration times, the analytical methods often become the bottleneck in the experimental workflow due to long analysis times. More rapid analytical methods could be employed that determine phase compositions spectroscopically (Carlsson and Karlberg, 2000; Marine et al., 2009; Thien et al., 2017, 2019), but they are limited to optically accessible components.

The laboriousness of LLE experiments has led to a lack of LLE data (Economou et al., 2014). In particular, multicomponent LLE are scarce: Currently, the Dortmund Data Bank (DortmundDataBank, 2019) contains LLE data on 13572 systems, of which 12606 are binary and ternary systems, 881 are quaternary systems, and only 85 are systems with more than 4 components.

2.2.2 Modeling liquid-liquid equilibria

A model-based analysis of LLE experiments requires a mathematical description of the experiment. In the following, the standard LLE model is briefly introduced. For details, the reader is referred to Gmehling et al. (2012) and Michelsen and Mollerup (2007).

Thermodynamically, two phases ' and " are in equilibrium, if they are in thermal, mechanical and chemical equilibrium:

$$T' = T'' \tag{2.37}$$

$$p' = p'' \tag{2.38}$$

$$\mu_i' = \mu_i'' \quad i = 1, ..., n_\mathrm{c}. \tag{2.39}$$

For LLE, Equation 2.39 can be rewritten in terms of mole fractions x_i and activity coefficients γ_i:

$$x_i'\gamma_i' = x_i''\gamma_i'' \quad i = 1, ..., n_\mathrm{c}. \tag{2.40}$$

The activity coefficients γ_i are typically calculated from Gibbs free energy (G^E) models, e.g., UNIQUAC (Abrams and Prausnitz, 1975) or NRTL (Renon and Prausnitz, 1968) (the UNIQUAC and NRTL model equations are given in Appendix E). However, in principle, equations of state can also be used.

The aim of the LLE model is to relate the phase compositions $\boldsymbol{x}'$ and $\boldsymbol{x}''$ to the overall composition $\boldsymbol{z}$. Thus, in addition to the equilibrium conditions, the LLE model has to include material balances. The total and the component mole balance read as

$$n_\mathrm{t} = n' + n'' \tag{2.41}$$

$$z_i n_\mathrm{t} = x_i' n' + x_i'' n'' \quad i = 1, ..., n_\mathrm{c} - 1. \tag{2.42}$$

Introducing the phase ratio $\beta = n'/n_\mathrm{t}$ of phase ', Equations 2.41-2.42 can be rewritten as

$$z_i = \beta x_i' - (1 - \beta) x_i'' \quad i = 1, ..., N_\mathrm{c} - 1. \tag{2.43}$$

Thus, Equation 2.43 describes the distribution of species i across the LLE phases. In addition, the compositions within each phase are related by the closing conditions

$$\sum_{i=1}^{N_\mathrm{c}} z_i = 1, \ \sum_{i=1}^{N_\mathrm{c}} x_i' = 1 \text{ and } \sum_{i=1}^{N_\mathrm{c}} x_i'' = 1. \tag{2.44}$$

Equations 2.44, 2.37, 2.38, 2.40, and 2.43 form the model for an LLE experiment. The entire model can be solved using the isothermal flash algorithm proposed by Michelsen and Mollerup (2007).

2.3 Contribution of this thesis

The discussion in the previous Sections 2.1 and 2.2 shows that a fast generation of accurate diffusion and LLE data is still a challenge and a major need in chemical engineering.

Regarding diffusion, 3 approaches are used to obtain diffusion coefficients:

- Diffusion experiments are the most significant source for diffusion coefficients and indispensable for validation purposes. However, diffusion experiments are burdened with high experimental effort and costs. In particular, multicomponent diffusion experiments are expensive. This circumstance has led to the situation that there is a lack of sufficient measurement data.
- The prediction of diffusion coefficients from MD simulations is versatile and offers detailed insight into molecular interactions. However, MD simulations depend on reliable force fields which are not always available. Furthermore, MD simulations can be computationally expensive.

- Engineering equations provide rapid estimations of diffusion coefficients that can be easily included in large scale simulations. However, engineering equations are not sufficiently accurate in many cases. In particular, predictions for non-ideal systems are burdened with large uncertainty. This is due to a lack of sufficient measurement data which is needed for the development and validation of engineering equations.

Regarding LLE, predictive methods are not yet sufficiently accurate. Therefore, LLE experiments are indispensable. However, LLE experiments are time-consuming and expensive. In particular, multicomponent LLE experiments are burdened with high costs. Therefore, LLE data are scarce.

Thus, there is a need for improved experiments and predictions that allow for a fast and accurate generation of diffusion and LLE data.

In this thesis, efficient diffusion and LLE experiments are developed and a new predictive equation for diffusion coefficients is derived. The improvements are developed by use of model-based experimental design and analysis methods which are introduced in Chapter 3.

A new predictive model for binary diffusion coefficients of non-ideal liquids

The progress in computational power and in MD simulation techniques offers new possibilities for faster predictions of diffusion coefficients from MD simulations. Thereby, MD simulations become attractive for the calculation of extensive data sets on pressure-, temperature-, and concentration-dependent diffusion coefficients. In addition, MD simulations offer a deep insight into the molecular interactions of a diffusion process. Thus, MD-simulated diffusion data is particularly suited for the development and validation of predictive equations for diffusion coefficients.

The discussion in Section 2.1.3 has shown that Darken-based engineering equations for the prediction of mutual diffusion coefficients suffer from the missing link between self-diffusion coefficients in pure substances and in the mixture. To provide the missing link, predictions of self-diffusion coefficients in mixtures are needed.

In Chapter 4, we develop a predictive model for self-diffusion coefficients in non-ideal binary liquid mixtures. The model is derived from MD simulations of Lennard-Jones systems and validated with experimental data from literature. Thus, our new model provides the missing link to render Darken-based models into practical tools to predict mutual diffusion coefficients.

Model-based optimal experimental design of diffusion experiments

The strive for improved diffusion experiments has produced a large variety of existing diffusion experiments. The accuracy of experimentally determined diffusion coefficients depends on the geometry of the experimental setup as well as on the analytical techniques used to determine concentrations of the diffusion species. Whereas analytical techniques and their accuracy are subject of continuous improvement, geometries are genuine features of a setup. In Chapter 5, we identify optimal geometries for diffusion experiments once and for all. In addition, optimal measurement conditions are derived, i.e. optimal measurement times and positions.

Experimental analysis of diffusion experiments

The development of new experiments is a difficult task. Both experimental setup and model have to be consistent to obtain statistically sound results. In Chapter 6, the application of model-based techniques for the development of improved diffusion experiments is demonstrated.

In Section 6.1, the model-based experimental approach is applied to overcome long-standing discrepancies in the measurement of diffusion in gases with a Loschmidt cell. Thereby, a consistent experimental setup for the measurement of diffusion in gases is developed that enables the measurement of concentration-dependent diffusion coefficients from a single experiment.

In Section 6.2, a microfluidic setup for the measurement of diffusion in liquids is developed that enables the measurement of multicomponent diffusion coefficients from a single experiment with minimum sample consumption.

Optimal experimental design of liquid-liquid equilibria experiments

Even by use of automated setups, LLE measurements are burdened with long measurement times. Therefore, a reduction of the *number* of experiments would be desirable.

In Chapter 7, optimal experimental designs for the characterization of LLE are identified. Thereby, a significant amount of experimental effort can be saved without losing accuracy of the LLE model parameters and predictions.

Optimal experimental design of diffusion and LLE experiments for extraction processes

Optimal experimental designs depend on the objective of investigation. For the measurement of physical properties, accurate model parameters and physical property predictions are of main interest. In contrast, industrial process applications do not strive for accurate physical properties, but for accurate predictions of the annual costs. Even though the accuracy of cost predictions depends on the accuracy of physical properties, process applications depend on a multitude of physical properties which can have significantly different impacts on costs.

Liquid-liquid extraction is a process that depends on both diffusion and LLE. Therefore, accurate cost estimations of extraction processes require accurate diffusion and LLE data. This leads to the fundamental questions: "Which data is more important for extraction processes: Diffusion or LLE? Which experiments should be performed in laboratory to obtain accurate cost estimations for extraction processes?" In Chapter 8, optimal designs for diffusion and LLE experiments are identified such that accurate cost estimations of extraction processes are obtained.

CHAPTER 3

Model-based design and analysis of experiments

The method of model-based design and analysis provides the basic tool for the investigations of this work. In the following, an overview on the theory required for the applications of this work is presented. It is based on the extensive and detailed treatises of Bard (1974) and Walter and Pronzato (1997).

3.1 Generic model representation

The application of model-based methods requires the existence of a mathematical model describing the experiment. A generic representation of such a model $\boldsymbol{g}$ is given in the *implicit* form

$$\boldsymbol{g}(\boldsymbol{w}, \boldsymbol{\theta}) = \mathbf{0}, \; \boldsymbol{g} \in \mathbb{R}^{N_\mathrm{g}}, \; \boldsymbol{w} \in \mathbb{R}^{N_\mathrm{w}}, \; \boldsymbol{\theta} \in \mathbb{R}^{N_\mathrm{P}} \tag{3.1}$$

Here, $\boldsymbol{w}$ are the variables of the model and $\boldsymbol{\theta}$ are the model parameters. The variables $\boldsymbol{w}$ can be split into independent variables $\boldsymbol{x}$, i.e. input variables, and dependent variables $\boldsymbol{y}$, i.e. output variables with:

$$\boldsymbol{w} = \{\boldsymbol{x}, \boldsymbol{y}\} \tag{3.2}$$

with $\dim(\boldsymbol{x}) = N_{\mathrm{x}} = N_{\mathrm{w}} - N_{\mathrm{g}}$ and $\dim(\boldsymbol{y}) = N_{\mathrm{y}} = N_{\mathrm{g}}$. Thus, the model $\boldsymbol{g}$ can be written in the form

$$\boldsymbol{g}(\boldsymbol{x}, \boldsymbol{y}, \boldsymbol{\theta}) = \boldsymbol{0}. \tag{3.3}$$

In many practical applications, Equation 3.3 can be reformulated into an *explicit* form

$$\boldsymbol{y} = f(\boldsymbol{x}, \boldsymbol{\theta}). \tag{3.4}$$

In principle, all elements of a model can be associated with uncertainty - the dependent variables $\boldsymbol{y}$, the independent variables $\boldsymbol{x}$, the model parameters $\boldsymbol{\theta}$ and even the model $\boldsymbol{g}$ itself. In an experiment, the independent variables $\boldsymbol{x}$ are set to specific values and the independent variables $\boldsymbol{y}$ are measured. For any measurement μ, the difference between the measured data $\boldsymbol{w}_{\mathrm{meas},\mu}$ and the true values $\boldsymbol{w}_{\mu}$ is called the *measurement error*

$$\boldsymbol{\epsilon}_{\mu} = \boldsymbol{w}_{\mathrm{meas},\mu} - \boldsymbol{w}_{\mu}. \tag{3.5}$$

For a statistically-based design and analysis of experiments, it is important to know the characteristics of the measurement error $\boldsymbol{\epsilon}_{\mu}$. In many practical applications and for the analyzes in this work, the following assumptions are appropriate with regard to the properties of the model and the measurement errors:

1. The model $\boldsymbol{g}$ is of exact-structural type, i.e., the model is assumed to apply exactly if the true values of the variables $\boldsymbol{w}$ and the model parameters $\boldsymbol{\theta}$ would be known. This assumption is often well justified when physical models are used that have been proofed to lead to consistent experimental results.

2. The model parameters $\boldsymbol{\theta}$ are identifiable, i.e., the model parameters $\boldsymbol{\theta}$ can be identified from the measurements.

3. There is no systematic error in the measurements.

4. The measurement errors in different experiments are independent and normally distributed with known constant covariance matrix $\boldsymbol{V}_{\boldsymbol{w}}$.

Whether or not the assumptions on the errors are appropriate has to be tested in the data analysis. In case of disagreements between the assumptions on the errors and the data analysis, an iterative refinement of the model and/or the experiment is required until agreement is achieved (cf. Section 3.3).

3.2 Model-based experimental design

An experimental design describes a specific combination of experiments. An experiment is characterized by its measurement conditions. In terms of the mathematical model $\boldsymbol{g}$ (Equation 3.3), the measurement conditions of an experiment μ correspond to the settings of the independent variables $\boldsymbol{x}_\mu$. In the context of experimental design theory, the independent variables $\boldsymbol{x}_\mu$ are also referred to as *design variables*. The design variables can be chosen within the design space Ω: $\boldsymbol{x}_\mu \in \Omega$. In practical applications, design variables can be, e.g., prepared sample compositions, temperatures, pressures, or measurement times. Mathematically, an experimental design is represented by the design vector

$$\boldsymbol{d} = \left\{ \begin{array}{cccc} \boldsymbol{x}_1 & \boldsymbol{x}_2 & \dots & \boldsymbol{x}_{n_\text{exp}} \\ v_1 & v_2 & \dots & v_{n_\text{exp}} \end{array} \right\}. \tag{3.6}$$

The design vector $\boldsymbol{d}$ combines the measurement conditions of the experiments, $\boldsymbol{x}_\mu$, with corresponding design weights, $v_\mu \in [0, 1]$, $\sum v_\mu = 1$. The design weights v_μ specify the *fraction of experimental effort* spent at experiment $\boldsymbol{x}_\mu$. n_exp denotes the number of *distinct experiments*, i.e. experiments with unique measurement conditions. Since experiments can be repeated, the number of distinct experiments n_exp is less or equal to the total number of experiments N_exp. If N_exp experiments are performed with each experiment repeated r_i times, the design weights are determined through $v_\mu = r_\mu / N_\text{exp}$ and are thus rational numbers. These kinds of designs are called *exact designs* and are denoted by $\boldsymbol{d}^{(N_\text{exp})}$.

When allowing the design weights to be real numbers, and increasing the total number of experiments $N_\text{exp} \to \infty$, $\boldsymbol{d}$ can be regarded as the normalized probability distribution of the experiments. This formulation leads to so-called *continuous designs*, which are denoted by $\boldsymbol{d}^*$ in the following.

3.2.1 Optimal experimental design

Experiments are always associated with measurement uncertainties. When experiments are evaluated, these measurement uncertainties lead to uncertainties in the estimated model parameters which in turn lead to uncertainties in the model predictions. Notably, the *strength* of this error propagation depends on the experimental design. Thus, optimal designs exist that lead to minimum uncertainties in the parameters and/or predictions.

For a given design $\boldsymbol{d}$, an approximate covariance matrix of the parameters, $\boldsymbol{V_\theta}$, can be calculated using the *Fisher information matrix* $\boldsymbol{F}$ (Bard, 1974):

$$\boldsymbol{V_\theta}(\hat{\boldsymbol{\theta}}, \boldsymbol{d}) \approx \left(N_{\text{exp}} \boldsymbol{F}(\hat{\boldsymbol{\theta}}, \boldsymbol{d})\right)^{-1}. \tag{3.7}$$

The parameter (co-)variance $\boldsymbol{V_\theta}$ is low if many experiments N_{exp} are performed and if the Fisher information $\boldsymbol{F}$ is high. The Fisher information $\boldsymbol{F}$ measures the sensitivity of the model with regard to the model parameters (Bard, 1974):

$$\boldsymbol{F}(\hat{\boldsymbol{\theta}}, \boldsymbol{d}) = \sum_{\mu=1}^{n_{\text{exp}}} v_\mu \boldsymbol{M}(\hat{\boldsymbol{\theta}}, \boldsymbol{w}_\mu) \; + \boldsymbol{\Sigma}_{\boldsymbol{0}}^{-1}, \tag{3.8}$$

where

$$\boldsymbol{M}(\hat{\boldsymbol{\theta}}, \boldsymbol{w}_\mu) = \boldsymbol{B}_\mu^{\mathrm{T}} \left(\boldsymbol{A}_\mu \boldsymbol{V}_{\boldsymbol{w}} \boldsymbol{A}_\mu^T\right)^{-1} \boldsymbol{B}_\mu \tag{3.9}$$

is the *local Fisher information* matrix with

$$\boldsymbol{A}_\mu = \left.\frac{\partial \boldsymbol{g}}{\partial \boldsymbol{w}}\right|_{\boldsymbol{w}_\mu, \hat{\boldsymbol{\theta}}} \quad \text{and} \quad \boldsymbol{B}_\mu = \left.\frac{\partial \boldsymbol{g}}{\partial \boldsymbol{\theta}}\right|_{\boldsymbol{w}_\mu, \hat{\boldsymbol{\theta}}}. \tag{3.10}$$

In equation 3.9, the term in brackets quantifies the uncertainty in the predictions of the model as a consequence of the measurement uncertainty by incorporating the measurement covariance matrix $\boldsymbol{V_w}$ and the sensitivity of the model equations $\boldsymbol{g}$ to the model variables $\boldsymbol{w}$. If the impact of measurement uncertainty is large, Fisher information is low. The matrix $\boldsymbol{B}$ expresses the sensitivity of the model equations on the unknown model parameters $\boldsymbol{\theta}$. Fisher information is high where the model output is highly sensitive to the parameter values. $\boldsymbol{\Sigma_0}$ is the (co-)variance-matrix of possibly available preliminary information on parameter uncertainty from previous experiments or simulations.

If only the independent variables $\boldsymbol{y}$ are subject to errors, Equation 3.9 reduces to

$$\boldsymbol{M}(\hat{\boldsymbol{\theta}}, \boldsymbol{w}_\mu) = \boldsymbol{B}_\mu^T \left(\boldsymbol{A}_\mu \boldsymbol{V}_{\boldsymbol{y}} \boldsymbol{A}_\mu^T\right)^{-1} \boldsymbol{B}_\mu \tag{3.11}$$

where

$$\tilde{\boldsymbol{A}}_\mu = \left.\frac{\partial \boldsymbol{g}}{\partial \boldsymbol{y}}\right|_{\boldsymbol{w}_\mu, \hat{\boldsymbol{\theta}}}. \tag{3.12}$$

For an explicit model $\boldsymbol{y} = f(\boldsymbol{x}, \boldsymbol{\theta})$, Equation 3.11 reduces further to

$$\boldsymbol{M}(\hat{\boldsymbol{\theta}}, \boldsymbol{w}_\mu) = \left(\frac{\partial \boldsymbol{y}}{\partial \boldsymbol{\theta}}\right)^{\mathrm{T}}_{\boldsymbol{w}_\mu, \hat{\boldsymbol{\theta}}} \boldsymbol{V}_{\boldsymbol{y}}^{-1} \left(\frac{\partial \boldsymbol{y}}{\partial \boldsymbol{\theta}}\right)_{\boldsymbol{w}_\mu, \hat{\boldsymbol{\theta}}}. \tag{3.13}$$

As can be seen from equation 3.8, the Fisher information matrix depends on an estimate $\hat{\boldsymbol{\theta}}$ of the unknown parameters (with linear models $\boldsymbol{g}$ being the only exception). This is a well-known dilemma of OED (Bard, 1974; Walter and Pronzato, 1997). However, in many cases, physical knowledge provides initial estimates $\hat{\boldsymbol{\theta}}$ which are already in a local neighborhood of the true values $\boldsymbol{\theta}$ (Bard, 1974; Schöneberger et al., 2010). If the estimates are not in a local neighborhood, the designed experiments based on those initial parameter estimates will not be optimal with respect to the real parameters. An iterative procedure can then be applied to alternate between the design and execution of experiments where the model parameters are refined after each execution of experiments (cf. Marquardt (2005); Walter and Pronzato (1997)). If required, robust optimal design methods could also be employed. Robust designs optimize, e.g., the expected value or the worst case of the uncertainty based on probability distributions for the estimated model parameters $\hat{\boldsymbol{\theta}}$ (Walter and Pronzato, 1997, 1987; Pronzato and Walter, 1988; Chu and Hahn, 2008; Barz et al., 2010).

Knowledge of the parameter covariances $\boldsymbol{V}_{\boldsymbol{\theta}}$ allows for the calculation of the uncertainty of the model predictions $\boldsymbol{y}$. The covariance matrix $\boldsymbol{V}_{\boldsymbol{y},\mathrm{Pred}}$ to be expected when the design $\boldsymbol{d}$ is executed is (Bard, 1974)

$$\boldsymbol{V}_{\boldsymbol{y},\mathrm{Pred}}(\boldsymbol{x}, \boldsymbol{d}, \hat{\boldsymbol{\theta}}) = \left(\tilde{\boldsymbol{A}}^{-1}\boldsymbol{B}\right) \boldsymbol{V}_{\boldsymbol{\theta}} \left(\tilde{\boldsymbol{A}}^{-1}\boldsymbol{B}\right)^{\mathrm{T}}. \tag{3.14}$$

In some cases, not the predictions of the dependent variables $\boldsymbol{y}$, but other specific functionals of the model parameters are of interest. These functionals could be, e.g., scalar cost functions $TAC(\boldsymbol{\theta})$ that depend on the model parameters. The variance $\sigma^2_{TAC,\mathrm{Pred}}$ in these cost functions follows from

$$\sigma^2_{TAC,\mathrm{Pred}}(\boldsymbol{d}, \hat{\boldsymbol{\theta}}) = \left(\frac{\partial TAC}{\partial \boldsymbol{\theta}}\right) \boldsymbol{V}_{\boldsymbol{\theta}} \left(\frac{\partial TAC}{\partial \boldsymbol{\theta}}\right)^{\mathrm{T}}. \tag{3.15}$$

The discussion above has shown that the uncertainties, i.e. the (co-)variances, in the model parameters and the model predictions depend on the experimental design. The aim of an *optimal* experimental design is to identify an experimental design minimizing the uncertainty in, e.g. the model parameters and/or the model predictions. Depending on the objective of the optimization, different objective functions Φ_i have been developed - the so-called "alphabetical design criterions" (Walter and Pronzato, 1997). Optimal designs $\boldsymbol{d}_{i,\text{opt}}$ are obtained from maximization of these objective functions Φ_i:

$$\boldsymbol{d}_{i,\text{opt}} = \underset{\boldsymbol{d}}{\operatorname{argmax}}\, \Phi_i. \tag{3.16}$$

In this work, the following objective functions are important:

- To determine the experimental design minimizing the uncertainty in the parameters $\boldsymbol{\theta}$, we use the popular *D-optimal design* to minimize the volume of the parameter confidence ellipsoid. For this purpose, the determinant of the Fisher information matrix $\boldsymbol{F}(\hat{\boldsymbol{\theta}}, \boldsymbol{d})$ is maximized (Bard, 1974). Thus, the objective function for a D-optimal design is

$$\Phi_\text{D} = \left(\det \boldsymbol{F}(\hat{\boldsymbol{\theta}}, \boldsymbol{d})\right)^{1/N_\text{p}}. \tag{3.17}$$

- To determine the experimental design leading to the most accurate model predictions of the dependent variables $\boldsymbol{y}$, we apply the *G-optimal design* which minimizes the maximum of the standardized prediction variance $\tilde{\sigma}_{\boldsymbol{y},\text{Pred}}$ (Atkinson et al., 2007). The corresponding objective function is

$$\Phi_\text{G} = \left(\max_{\boldsymbol{x}} \tilde{\sigma}_{\boldsymbol{y},\text{Pred}}(\boldsymbol{x}, \boldsymbol{d}, \hat{\boldsymbol{\theta}})\right)^{-1}. \tag{3.18}$$

  Here, $\tilde{\sigma}_{\boldsymbol{y},\text{Pred}}$ is the standardized prediction variance which is a normalized, scalar measure of $\boldsymbol{V}_{\boldsymbol{y},\text{Pred}}$ at any independent variable value $\boldsymbol{x}$:

$$\tilde{\sigma}_{\boldsymbol{y},\text{Pred}}(\boldsymbol{x}, \boldsymbol{d}, \hat{\boldsymbol{\theta}}) = \frac{1}{N_\text{exp}} \sum_u^{N_y} \sum_v^{N_y} \sigma^{uv} V_{\boldsymbol{y},\text{Pred},uv}(\boldsymbol{x}, \boldsymbol{d}, \hat{\boldsymbol{\theta}}), \tag{3.19}$$

  where σ^{uv} is the u-v-th element of $\boldsymbol{V}_{\boldsymbol{y}}^{-1}$ and $V_{\boldsymbol{y},\text{Pred},uv}$ is the u-v-th element of $\boldsymbol{V}_{\boldsymbol{y},\text{Pred}}$ (cf. Equation 3.14). Thus, σ_Pred is an average value of the variance in the model predictions $\boldsymbol{y}$. Therefore, the G-optimal design is a min-max-design that determines the experimental settings $\boldsymbol{d}$ that minimize the mean variance in the model predictions $\boldsymbol{y}$ at the worst dependent variable value $\boldsymbol{x}$.

- To determine the experimental design minimizing the uncertainty in a scalar cost function $TAC(\boldsymbol{\theta})$ that depends on the model parameters, we apply the *c-optimal design* (Walter and Pronzato, 1997) which maximizes

$$\Phi_{\mathrm{c}} = \left(\sigma^2_{TAC,\mathrm{Pred}}(\boldsymbol{d}, \hat{\boldsymbol{\theta}})\right)^{-1} \tag{3.20}$$

where $\sigma^2_{TAC,\mathrm{Pred}}(\boldsymbol{d}, \hat{\boldsymbol{\theta}})$ follows from Equation 3.15.

Both D- and G-optimal designs are interlinked through the *General Equivalence Theorem* (GET) (Walter and Pronzato, 1997; Kiefer and Wolfowitz, 1960; Atkinson et al., 2007). GET states in the continuous design case (a) that G- and D-optimal designs are equal and (b) that the maximum of the standardized prediction variance is equal to the number of parameters N_{p}:

$$\boldsymbol{d}^*_{\mathrm{G}} = \boldsymbol{d}^*_{\mathrm{D}} \tag{3.21}$$

$$\max_{\boldsymbol{x}} \tilde{\sigma}_{\boldsymbol{y},\mathrm{Pred}}(\boldsymbol{x}, \boldsymbol{d}^*_{D}) = N_{\mathrm{p}}. \tag{3.22}$$

For any design criterion Φ_i, the quality of an experimental design $\boldsymbol{d}$ can be evaluated by the efficiency

$$\zeta_i(\boldsymbol{d}) = \frac{\Phi_i(\boldsymbol{d})}{\Phi_i(\boldsymbol{d}^*_{\mathrm{opt}})}. \tag{3.23}$$

The efficiency ζ_i is a measures for the information loss when using the design $\boldsymbol{d}$ instead of $\boldsymbol{d}^*_{\mathrm{opt}}$, i.e. $\zeta_i = 0$ denotes entire information loss whereas $\zeta_i = 1$ implies no information loss. The efficiency is defined such that an experiment with a value of $\zeta_i = 0.5$ would have to be conducted $\frac{1}{\zeta_i} = 2$ times to obtain the same accuracy as in the optimal continuous experiment.

Continuous D- and c-optimal designs can be constructed sequentially (Titterington, 1976; Silvey et al., 1978; Yu et al., 2010). For this purpose, the design space is discretized and an initial design vector $\boldsymbol{d}^{(0)}$ with arbitrary, non-zero design weights $v_i^{(0)}$ is defined. In an iterative procedure with iterative index $q = 0, 1, ...$, the design weights $v_i^{(q)}$ are then updated using the following algorithm:

$$v_i^{(q+1)} = v_i^{(q)} \frac{h_i^\lambda}{\sum_{j=1}^{n_{\exp}} v_i^{(q)} h_j^\lambda}, \qquad i = 1, ..., n_{\exp} \tag{3.24}$$

where $\lambda \in (0, 1]$ is a calibration parameter. The function h_i is given by

$$h_i = \mathrm{tr}\left(\Phi'\left(\boldsymbol{F}(\boldsymbol{d}^q)\right) \boldsymbol{M}(\boldsymbol{x}_i)\right) \tag{3.25}$$

where $\Phi'(\boldsymbol{F}) = \partial\Phi(\boldsymbol{F})/\partial\boldsymbol{F}$ is the gradient of the objective design function (cf. Equations 3.17-3.20) and $\boldsymbol{M}$ is the local Fisher information matrix (cf. Equation 3.9). Yu et al. (2010) proved that under only mild conditions the algorithm Equations 3.24-3.25 converge monotonically to the optimal design $\boldsymbol{d}_{\mathrm{opt}}^*$ of the corresponding objective function Φ for the D- or c-optimal designs.

In addition to the algorithm Equations 3.24-3.25, an appropriate stopping criterion is needed. Holland-Letz et al. (2012) has shown that a lower bound on the efficiency ζ_i is given by

$$\zeta_i(\boldsymbol{d}) \geq \left(\max_{\boldsymbol{x}} \frac{\mathrm{tr}\left(\Phi'\left(\boldsymbol{F}(\boldsymbol{d})\right) \boldsymbol{M}(\boldsymbol{x}_i)\right)}{\mathrm{tr}\left(\Phi'\left(\boldsymbol{F}(\boldsymbol{d})\right) \boldsymbol{F}(\boldsymbol{d})\right)} \right). \tag{3.26}$$

Thus, Equation 3.26 can be used to evaluate the progress in the numerical construction of a continuous optimal experimental design.

G-optimal continuous designs can be obtained from D-optimal continuous designs according to the GET (cf. Equations 3.21 and 3.22).

For the calculation of exact experimental designs, specific algorithms are needed in general. However, as a good approximation, the design weights of a continuous design can be rounded to match the discrete number of experiments in an exact design.

3.3 Model-based experimental evaluation and parameter estimation

After the execution of experiments, the experimental data has to be analyzed using appropriate data analysis methods. The appropriateness of a data analysis method is particularly determined by two factors:

1. The selection of an appropriate error model: The error model has to reflect the real errors in the experiment and/or the model.
2. The appropriate interpretation of the data analysis results: A detailed analysis of the data analysis is indispensable to verify if the model is adequate to describe the experiment. If required, the model and/or the experiment have to be refined to obtain statistically sound results.

The error model used in this work is introduced in Section 3.1. In particular, we assume an exact-structural model $\boldsymbol{g}(\boldsymbol{w}, \boldsymbol{\theta}) = \mathbf{0}$. The measurement errors are assumed to be free of systematic errors and to be normally distributed with constant covariance matrix $\boldsymbol{V}_{\boldsymbol{w}}$.

The aim of parameter estimation is to determine the unknown model parameters $\boldsymbol{\theta}$ of the model based on the measurement data $\boldsymbol{w}_{\text{meas}}$. Due to measurement errors $\boldsymbol{\epsilon}_\mu = \boldsymbol{w}_{\text{meas},\mu} - \boldsymbol{w}_\mu$, the measurement data $\boldsymbol{w}_{\text{meas},\mu}$ differs from the true values $\boldsymbol{w}_\mu$ for every measurement μ (cf. Equation 3.5). Since the true values $\boldsymbol{w}_\mu$ are unknown, the measurement error $\boldsymbol{\epsilon}_\mu$ cannot be known. However, since we assume an exact structural model $\boldsymbol{g}$, the *residuals*

$$\boldsymbol{e}_\mu = \boldsymbol{w}_{\text{meas},\mu} - \hat{\boldsymbol{w}}_\mu \tag{3.27}$$

can be calculated as a prediction of the measurement errors. Here, $\hat{\boldsymbol{w}}_\mu$ are the variable values that satisfy the model $\boldsymbol{g}(\hat{\boldsymbol{w}}_\mu, \hat{\boldsymbol{\theta}})$ for a parameter estimate $\hat{\boldsymbol{\theta}}$. If the parameter estimate $\hat{\boldsymbol{\theta}}$ is close to the true parameters $\boldsymbol{\theta}$, the residuals should in turn be close to the true measurement errors. As a consequence, the statistical properties of the residuals should be close to the statistical properties of the measurement errors.

Vice versa, the parameters $\hat{\boldsymbol{\theta}}$ can be estimated such that the statistical properties of the residuals are as close to the statistical properties of the measurement errors as possible with *maximum likelihood.* This is the idea of the maximum likelihood parameter estimation procedure.

The maximum likelihood parameters are obtained from the optimization problem

$$\hat{\boldsymbol{\theta}} = \underset{\boldsymbol{\theta}}{\operatorname{argmin}}\, \hat{\Phi}_i \tag{3.28}$$

Under the assumptions on the model and the error model made in this work (cf. Section 3.1), the objective function $\hat{\Phi}_i$ reduces to the well known weighted least-squares criterion

$$\Phi_{\text{WLS}} = \sum_{\mu=1}^{N_{\text{exp}}} \boldsymbol{e}_{\mu}^{\text{T}} \boldsymbol{V}_{\boldsymbol{w}}^{-1} \boldsymbol{e}_{\mu}. \tag{3.29}$$

If the covariance matrix has the form $\boldsymbol{V}_{\boldsymbol{w}} = \sigma_w \boldsymbol{I}$, i.e. errors in different variables w_i are independent of each other and have the same uncertainty σ_w, Equation 3.29 reduces to the least-squares criterion

$$\Phi_{\text{LS}} = \sum_{\mu=1}^{N_{\text{exp}}} \boldsymbol{e}_{\mu}^{\text{T}} \boldsymbol{e}_{\mu}. \tag{3.30}$$

The first step after computation of parameter estimates should be the analysis of the residuals. For the error model assumed in this work, the residuals should be normally distributed. Therefore, the residuals should be free of drifts in time and space. A visual examination of the residuals can give a first hint on the quality of the parameter estimation. In addition, an estimate for the covariance matrix of the residuals should be computed:

$$\hat{\boldsymbol{V}} = \frac{N_r/N_g}{N_d - N_p/N_g} \sum_{\mu=1}^{N_{\text{exp}}} \boldsymbol{e}_{\mu} \boldsymbol{e}_{\mu}^{\text{T}}. \tag{3.31}$$

Here, N_r is the number of variables subject to error, i.e. the number of dependent and/or independent variables. N_d is the number of measured datapoints, i.e. $N_d = N_{\text{exp}} N_y$. For a single equation model ($N_g = 1$) with only the independent variable subject to error ($N_r = 1$), Equation 3.31 reduces to the well known formula for the root-mean-square error $RMSE$

$$RMSE = \sqrt{\frac{1}{N_d - N_p} \sum_{\mu=1}^{n_{\text{exp}}} \boldsymbol{e}_{\mu}^2}. \tag{3.32}$$

In a second step, the estimated covariance matrix of the parameters should be computed (cf. Section 3.2):

$$\boldsymbol{V}_{\boldsymbol{\theta}} = \left(\sum_{\mu=1}^{N_{\text{exp}}} \left(\frac{\partial \boldsymbol{w}}{\partial \boldsymbol{\theta}} \right)_{\boldsymbol{w}_{\mu}, \hat{\boldsymbol{\theta}}}^{\text{T}} \hat{\boldsymbol{V}}^{-1} \left(\frac{\partial \boldsymbol{w}}{\partial \boldsymbol{\theta}} \right)_{\boldsymbol{w}_{\mu}, \hat{\boldsymbol{\theta}}} \right)^{-1}. \tag{3.33}$$

The parameter covariance matrix allows for the analysis of correlations between different parameters θ_i and θ_j. For this purpose, correlation coefficients ρ_{ij} can be computed:

$$\rho_{ij} = \frac{V_{\boldsymbol{\theta},ij}}{\sqrt{V_{\boldsymbol{\theta},ii} V_{\boldsymbol{\theta},jj}}}, \quad -1 \leq \rho_{ij} \leq 1. \tag{3.34}$$

If the absolute value of a correlation parameter $\rho_{ij}$is close to 1, parameters θ_i and θ_j are highly correlated; the estimation of θ_i and θ_j is difficult. In addition, chances are high that the model $\boldsymbol{g}$ includes too many parameters. Thus, the risk of overfitting is high.

The parameter covariance matrix does not always give a vivid description of the parameter uncertainty. Parameter confidence intervals can be a helpful complement. Parameter confidence intervals are obtained from the confidence ellipsoid

$$(\boldsymbol{\theta} - \hat{\boldsymbol{\theta}})^{\mathrm{T}} V_{\boldsymbol{\theta}}^{-1} (\boldsymbol{\theta} - \hat{\boldsymbol{\theta}}) = N_p F_{1-\alpha}(N_p, N_d - N_p), \tag{3.35}$$

around the estimated parameters $\hat{\boldsymbol{\theta}}$. $F_{1-\alpha}$ is the F-distribution for the statistical significance α.

If (1) the residuals match the assumed error model, (2) the parameters are low correlated, and (3) the parameter uncertainties are acceptable, the results of the model-based experimental analysis are of statistical significance. Otherwise, a refinement of the model and/or experiment is required.

To sum up, the model-based experimental data analysis includes the following steps:

1. Setup of a model $\boldsymbol{g}(\boldsymbol{w}, \boldsymbol{\theta})$ and an error model.
2. Execution of experiments and generation of measurement data $\boldsymbol{w}$.
3. Calculation of the parameter estimates $\hat{\boldsymbol{\theta}}$.
4. Analysis of the statistical properties of residuals and the parameter estimates. If disagreements with the assumed error model and/or the required parameter uncertainty are observed, a revision of the model, the error model, and/or the experiment is required. Eventually, nuisance parameters have to be introduced into the model. However, overfitting has to be avoided. After the refinement of the model and/or experiment, a reiteration to step 2 should be performed until satisfying results are obtained.

Chapter 4

A new predictive model for binary diffusion coefficients of nonideal liquids

Reprinted (adapted) with permission from
Wolff, L., Jamali, S. H., Becker, T. M., Moultos, O. A., Vlugt, T. J. H., and Bardow, A. (2018). Prediction of composition-dependent self-diffusion coefficients in binary liquid mixtures: The missing link for Darken-based models, *Ind. Eng. Chem. Res*, 57:14784-14794. (Wolff et al., 2018a), Copyright (2018) American Chemical Society,
http://pubs.acs.org/articlesonrequest/AOR-NSu7bFvBu4NFtmTkS8R3,
Contribution of the author: Writing the draft, principal author, analysis of the raw data derived from MD simulations conducted by S. H. Jamali, development of the diffusion model.

In Chapter 2, the need for predictive models for diffusion coefficients was emphasized. Existing predictive diffusion models work well for approximately ideal mixtures, but there is a need for predictive diffusion models for non-ideal mixtures. However, a lack of sufficient diffusion data hinders the identification of a suitable model structure for the prediction of diffusion coefficients.

Molecular Dynamics (MD) simulations are a powerful tool to complement or even substitute diffusion experiments (Liu et al., 2013; Guevara-Carrion et al., 2016a,b). MD simulations do not only provide mutual diffusion coefficients, but a full set of transport data and thermodynamic properties, e.g. correlation functions and thermodynamic factors. Therefore, MD simulations can give detailed insight into diffusion processes not accessible in diffusion experiments. Running a large number of MD simulations can provide an extensive and consistent dataset which can be used as a basis for the development of predictive diffusion models.

The existing predictive diffusion models predict the *mutual* diffusion coefficients as a function of the *self*-diffusion coefficients. The successful performance of the Moggridge equation (Equation 2.33) shows that the inclusion of a function of the thermodynamic factor Γ into the ideal *mutual* diffusion equation (the Darken Equation 2.32) can be a sufficient method to take nonidealities into account. The question arises whether it is also possible to correct the ideal *self*-diffusion equation (the McCarty-Mason Equation 2.36) with a function of the thermodynamic factor Γ to expand its applicability to non-ideal mixtures.

In this chapter, we study the concentration-dependence of mutual and self-diffusion coefficients in binary non-ideal liquid mixtures. We investigate the correlation of non-ideal diffusion effects with the thermodynamic factor. To have a full and consistent set of transport data and thermodynamic data, we use MD simulations of Lennard-Jones systems for our analysis. In Section 4.1, we motivate our analysis from a theoretical point of view. In Section A.1, we present the results of our analysis: In a first step, we assess the performance of the Moggridge equation (Equation 2.33) for the prediction of mutual diffusion coefficients (Section 4.2.1). Subsequently, we analyze the non-ideal behavior of self-diffusion coefficients and derive an improved model for the prediction of self-diffusion coefficients in strongly non-ideal binary liquid mixtures (Section 4.2.2). The improved model is tested and validated with experimental data of molecular systems (Section 4.2.2). Conclusions of this study are drawn in Section 4.3.

4.1 Theory and method

The mutual MS diffusion coefficient $Đ_{12}$ can be split into two parts: An (ideal) Darken diffusion coefficient $Đ_{\text{Darken}}$ containing the self-diffusion coefficients $D_{i,\text{self}}$ (which are velocity autocorrelations), and a non-ideal diffusion coefficient $Đ_{\text{Cross}}$ containing the velocity crosscorrelations CC_{ij} between different particles of components i and j

(cf. Equation 2.27):

$$Đ_{12} = \underbrace{x_2 D_{1,\text{self}} + x_1 D_{2,\text{self}}}_{Đ_{\text{Darken}}} + \underbrace{x_1 x_2 N (CC_{11} + CC_{22} - 2CC_{12})}_{Đ_{\text{Cross}}}. \tag{4.1}$$

The nonideality of a mixture can be represented by the thermodynamic factor Γ. The thermodynamic factor Γ can be calculated from excess enthalpy models (Taylor and Krishna, 1993) or equations of state (Krishna and van Baten, 2005) (cf. Section 2.1.3). In molecular simulations, the thermodynamic factor can be calculated from Kirkwood-Buff integrals G_{ij} (Krüger et al., 2013; Liu et al., 2013):

$$\Gamma = \left(1 + \frac{1}{V} x_1 x_2 N \left(G_{11} + G_{22} - 2G_{12}\right)\right)^{-1}. \tag{4.2}$$

Here, V is the volume of the simulation box and N is the total number of particles. For details, the reader is referred to Milzetti et al. (2018), Ben-Naim (2006), and Jamali et al. (2018)

A number of works have studied the concentration dependence of velocity cross-correlations CC_{ij} (Weingärtner and Bertagnolli, 1986; Weingärtner, 1990; Marbach et al., 1995; Weingärtner, 2005; Hertz et al., 1977; Hertz and Mills, 1978; Mills and Hertz, 1980; Friedman and Mills, 1981; Geiger et al., 1981; Friedman and Mills, 1986; Easteal et al., 1987; Mills and Woolf, 1992). Weingärtner (1990) observed that velocity crosscorrelations CC_{ij} show a similar concentration-dependence as Kirkwood-Buff coefficients G_{ij}. However, a derivation of a relationship between velocity crosscorrelations CC_{ij} and the thermodynamic factor Γ is not straight-forward and no conclusive answer was found. Still, it is interesting to note that Weingärtner's observation in fact suggests a connection between the non-ideal diffusion coefficient $Đ_{\text{Cross}}$ and the thermodynamic factor Γ: A comparison of Equation 4.2 and Equation 4.1 reveals a structural similarity in the formulations of $Đ_{\text{Cross}}$ and Γ. Similarly, the Moggridge equation (Equation 2.33) also suggests a correlation between the relative nonideality $Đ_{\text{Cross}}/Đ_{\text{Darken}}$ and Γ: insertion of the Moggridge equation (Equation 2.33) into Equation 4.1 leads to

$$\frac{Đ_{\text{Cross}}}{Đ_{\text{Darken}}} = \frac{Đ_{12} - Đ_{\text{Darken}}}{Đ_{\text{Darken}}} \tag{4.3}$$

$$= \frac{Đ_{12}}{Đ_{\text{Darken}}} - 1 \qquad (4.4)$$

$$= \Gamma^{-0.36} - 1. \qquad (4.5)$$

Thus, the Moggridge equation (Equation 2.33) in fact relates the non-ideal diffusion coefficient $Đ_{\text{Cross}}$ to the ideal Darken diffusion coefficient $Đ_{\text{Darken}}$ and the thermodynamic factor Γ.

The question arises whether the relation between non-ideal diffusion effects and the thermodynamic factor can also be observed for self-diffusion coefficients. The McCarty-Mason equation (Equation 2.36) resembles the ideal mixing rule for concentration-dependent self-diffusion coefficients. In an analogy to the relative non-ideality $Đ_{\text{Cross}}/Đ_{\text{Darken}}$ of the mutual diffusion coefficient (Equation 4.3), we define the relative deviation $\Delta D_{i,\text{self,rel}}$ between the real self-diffusion coefficient $D_{i,\text{self}}$ and the self-diffusion coefficient $D_{i,\text{self,pred}}$ predicted by the McCarty-Mason equation (Equation 2.36),

$$\Delta D_{i,\text{self,rel}} = \frac{D_{i,\text{self}} - D_{i,\text{self,pred}}}{D_{i,\text{self}}}, \qquad (4.6)$$

which is a measure for non-ideal effects of self-diffusion. If the relative deviation $\Delta D_{i,\text{self,rel}}$ can be described as a function $f(\Gamma)$ of the thermodynamic factor, predictions of the McCarty-Mason equation (Equation 2.36) can be corrected to obtain a predictive equation for non-ideal mixtures:

$$\Delta D_{i,\text{self,rel}} = \frac{D_{i,\text{self}} - D_{i,\text{self,pred}}}{D_{i,\text{self}}} = f(\Gamma), \qquad (4.7)$$

$$\frac{1}{D_{i,\text{self}}} = \frac{1}{D_{i,\text{self,pred}}} \cdot (1 - f(\Gamma)) \qquad (4.8)$$

$$= \left(\frac{x_1}{D_{i,\text{self}}^{x_1 \to 1}} + \frac{x_2}{D_{i,\text{self}}^{x_2 \to 1}} \right) \cdot (1 - f(\Gamma)). \qquad (4.9)$$

In the following Section 4.2, we investigate correlations between both the relative nonideality $Đ_{\text{Cross}}/Đ_{\text{Darken}}$ of mutual diffusion coefficients and the relative deviation $\Delta D_{i,\text{self,rel}}$ of self-diffusion coefficients with the thermodynamic factor Γ. We expect the thermodynamic factor to cover all nonidealities such that there is no need for further correction factors such as viscosity. To have a full and consistent set of transport data and thermodynamic data, we use MD simulations of Lennard-Jones (LJ) systems for our analysis. The MD simulations have been performed by S. H. Jamali in the framework of another work of the present author and co-workers (Jamali et al., 2018). Details on the specifications of the MD simulation can be found in Appendix A.1. The identified correlations are then tested with experimental data of molecular systems.

4.2 Results and discussion

The MD simulations provide a full set of transport data and thermodynamic properties. Thereby, the MD simulations enable a comprehensive analysis of non-ideal effects of mutual and self-diffusion coefficients. In Section 4.2.1, we analyze the correlation between the relative nonideality $Đ_{\mathrm{Cross}}/Đ_{\mathrm{Darken}}$ of mutual diffusion coefficients and the thermodynamic factor Γ. We assess the performance of the Moggridge equation (Equation 2.33) and confirm its validity for a wide range of non-ideal mixtures. In Section 4.2.2, we investigate the correlation between the relative deviation $\Delta D_{i,\mathrm{self,rel}}$ of self-diffusion coefficients with the thermodynamic factor Γ. We derive an improved model for the prediction of concentration-dependent self-diffusion coefficients in non-ideal binary mixtures (Section 4.2.2) and validate our model with experimental data (Section 4.2.2).

4.2.1 Mutual diffusion coefficients

The nonideality of mutual MS diffusion coefficients $Đ_{12}$ is represented by the relative nonideality $Đ_{\mathrm{Cross}}/Đ_{\mathrm{Darken}}$ (Equation 4.3). According to Equation 4.5, we can assume a correlation between the relative nonideality $Đ_{\mathrm{Cross}}/Đ_{\mathrm{Darken}}$ and the thermodynamic factor Γ. Figure 4.1 shows the relative nonideality $Đ_{\mathrm{Cross}}/Đ_{\mathrm{Darken}}$ as a function of the thermodynamic factor Γ. The data of our MD simulations are a continuous function of the thermodynamic factor Γ. For the considered LJ systems, the thermodynamic factor is in the range $0.28 < \Gamma < 9$ and the relative nonidealities are in the range $-0.34 < Đ_{\mathrm{Cross}}/Đ_{\mathrm{Darken}} < 0.47$, i.e. the MS diffusion coefficient $Đ_{12}$ differs from the ideal Darken diffusion coefficient by up to 47 %. For ideal mixtures with weak molecular interactions ($\Gamma = 1$), the velocity crosscorrelations vanish and thereby the relative nonideality vanishes: $Đ_{\mathrm{Cross}}/Đ_{\mathrm{Darken}} = 0$. For mixtures with strong attractive interactions between particles of the same component i ($\Gamma < 1$), the velocity crosscorrelations CC_{ii} become predominant and the relative nonideality $Đ_{\text{Cross}}/Đ_{\text{Darken}}$ is positive. For mixtures with strong attractive interactions between unlike particles of different components i and j, the velocity croscorrelations CC_{ij} become predominant and the relative nonideality $Đ_{\text{Cross}}/Đ_{\text{Darken}}$ is negative.

To validate our MD simulation data, we compare it to experimental data from literature (a detailed list of references for the experimental data is provided in Appendix A.2). The experimental datasets consist of mutual Fick diffusion coefficients D_{12}, self-diffusion coefficients $D_{i,\text{self}}$, and thermodynamic factors Γ. The self-diffusion coefficients $D_{i,\text{self}}$ originate from NMR measurements or diffusion measurements with radioactive tracers; the thermodynamic factors Γ are either reported directly in literature or calculated from Redlich-Kister (RK) and/or NRTL parameters reported in literature. The relative nonideality $Đ_{\text{Cross}}/Đ_{\text{Darken}}$ is calculated from the experimental datasets via combination of Eqs. 4.1 and 2.12:

$$Đ_{\text{Cross}}/Đ_{\text{Darken}} = \frac{Đ_{12} - Đ_{\text{Darken}}}{Đ_{\text{Darken}}} \tag{4.10}$$

$$= \frac{D_{12}/\Gamma}{Đ_{\text{Darken}}} - 1 \tag{4.11}$$

$$= \frac{D_{12}/\Gamma}{x_2 D_{1,\text{self}} + x_1 D_{2,\text{self}}} - 1. \tag{4.12}$$

The thermodynamic factors of the experimental data are in the range $0 < \Gamma < 2$. Figure 4.1 provides an inset for the range $0 < \Gamma < 2$. Overall, our MD data agree well with the experimental data. Deviations can be observed only for mixtures with dimerising species, i.e., ethanol and methanol for the current dataset. For dimerising species, the relative nonideality $Đ_{\text{Cross}}/Đ_{\text{Darken}}$ is larger in comparison to non-dimerising species. This special behavior of dimerising species was also observed by Moggridge (2012a) and is also observed for self-diffusion coefficients (cf. Section 4.2.2).

Figure 4.1 also shows the predictions of the Moggridge equation (cf. Equation 2.33 and Equation 4.5). In the typical range of thermodynamic factors of molecular systems, $0 < \Gamma < 2$, the Moggridge equation performs well and agrees with our MD data as well as with most of the experimental data. Again, mixtures with dimerising species show larger deviations from the Moggridge equation. For large thermodynamic factors $\Gamma > 2$, our MD data suggests a different functional relation than the Moggridge equation with less negative relative nonidealities $Đ_{\text{Cross}}/Đ_{\text{Darken}}$. However, for practical applications with typical thermodynamic factors $0 < \Gamma < 2$, the performance of the Moggridge equation is excellent.

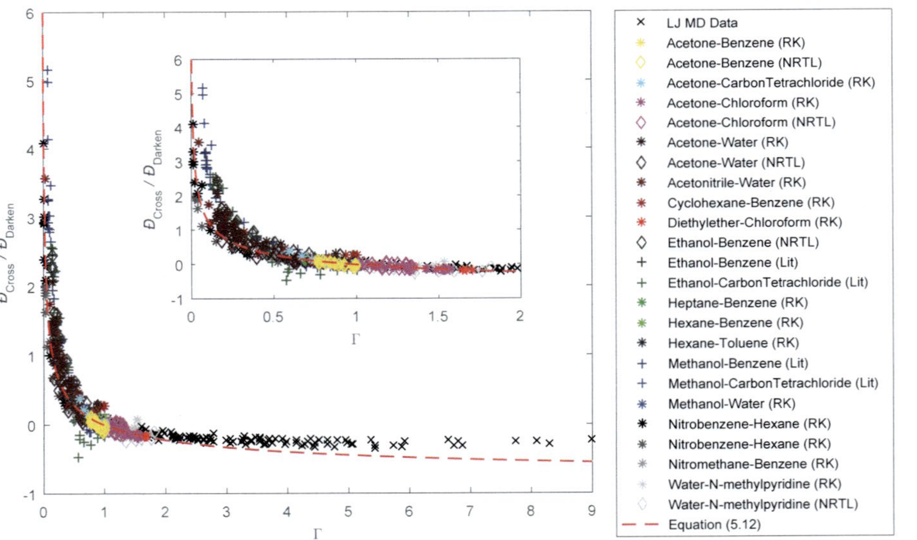

Figure 4.1: Relative nonideality $Đ_{\text{Cross}}/Đ_{\text{Darken}}$ of mutual diffusion coefficients as function of the thermodynamic factor Γ. Inset for thermodynamic factors $0 < \Gamma < 2$. Black crosses: Data from MD simualations. Stars: Experimental data with thermodynamic factors calculated with Redlich-Kister (RK). Diamonds: Experimental data with thermodynamic factors calculated with NRTL. Plus symbols: Experimental data with thermodynamic factors given in literature. Red dashed line: Predictive Moggridge equation (Equation 2.33). Statistical uncertainties of the MD data are given in Jamali et al. (2018). References for the experimental data are provided in Table A.2.

4.2.2 Self-diffusion coefficients

The nonideality of self-diffusion coefficients $D_{i,\text{self}}$ is represented by the relative deviation $\Delta D_{i,\text{self,rel}}$ (Equation 4.6). For ideal mixtures, we expect relative deviations $\Delta D_{i,\text{self,rel}} = 0$ from the predictions of the McCarty-Mason equation (Equation 2.36). For non-ideal mixtures, we assume that a modification of the McCarty-Mason equation with a function of the thermodynamic factor can account for non-ideal effects (cf. Equation 4.9). In Section 4.2.2, we analyze the correlation between the relative deviation $\Delta D_{i,\text{self,rel}}$ and the thermodynamic factor Γ in non-ideal LJ systems and derive a modified McCarty-Mason equation for non-ideal mixtures. Subsequently, we validate the modified McCarty-Mason equation with experimental data.

Self-diffusion coefficients of LJ systems

Figure 4.2 (top figure) shows an example of a concentration-dependent self-diffusion coefficient $D_{1,\text{self}}$ of a binary LJ system with pronounced nonideality. The specification of the LJ system is $\epsilon_2/\epsilon_1 = 0.6$, $\sigma_2/\sigma_1 = 1.2$, $m_2/m_1 = 1.728$, $k_{ij} = -0.6$.

In a first step, we test the performance of the McCarty-Mason equation (Equation 2.36). The McCarty-Mason equation requires the prior knowledge of self-diffusion coefficients $D_{i,\text{self}}^{x_j \to 1}$ at infinite dilution and of the pure substances. In MD simulations, statistical uncertainties are very large for mixtures approaching infinite dilution of one of the components. Therefore, the MD simulations have been performed for mixtures with at least 10 mol % of each species, i.e. $x_1 = [0.1, 0.3, 0.5, 0.7, 0.9]$. To obtain the values of $D_{1,\text{self}}^{x_j \to 1}$, we performed a smoothing fit with a quadratic polynomial function to the self-diffusion coefficients $D_{1,\text{self}}$. Figure 4.2 (top figure) shows the smoothing fit as well as the predictions of the McCarty-Mason equation. As expected, the McCarty-Mason prediction shows large deviations. However, the curvature of the concentration-dependence is retrieved.

Figure 4.2 (bottom figure) shows the concentration dependence of the relative deviation $\Delta D_{1,\text{self,rel}}$ (Equation 4.6) of the predictions made by the McCarty-Mason equation. Large relative deviations up to 70 % are observed. Figure 4.2 (bottom figure) also shows the concentration dependence of the thermodynamic factor minus 1, $(\Gamma - 1)$. The term $(\Gamma - 1)$ is a measure for the deviation of the mixture from an ideal mixture. In the present case, $(\Gamma - 1)$ takes values of up to 4.2, i.e. the LJ system is

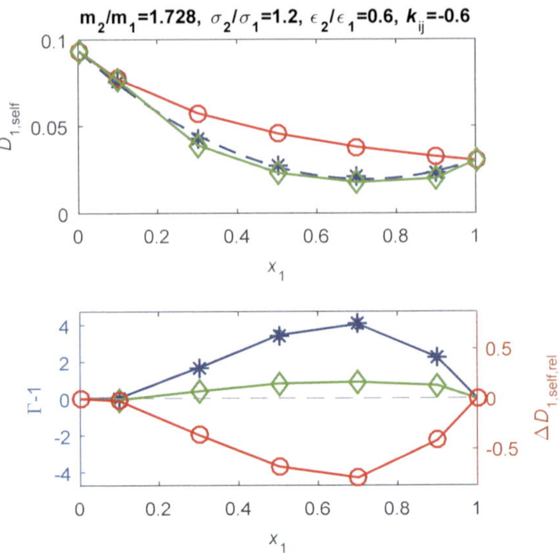

Figure 4.2: **Top figure**: Blue stars: Simulation results of self-diffusion coefficients $D_{1,\mathrm{self}}$ of a binary LJ system as function of the mole fraction x_1 of the first species. Specification of the LJ system: $\epsilon_2/\epsilon_1 = 0.6$, $\sigma_2/\sigma_1 = 1.2$, $m_2/m_1 = 1.728$, $k_{ij} = -0.6$. Blue-dashed line: smoothing fit to the simulation results; red circles/line: predictions of the McCarty-Mason equation (Equation 2.36); green diamonds/line: predictions of the modified McCarty-Mason equation (Equation 4.14). The error bars of $D_{1,\mathrm{self}}$ are smaller than the symbols.
Bottom figure: Concentration dependence of the thermodynamic factor $\Gamma - 1$ (blue stars/line, left axis) and concentration-dependence of the relative deviation $\Delta D_{i,\mathrm{self,rel}}$ between the self-diffusion coefficients and the McCarty-Mason predictions (Equation 2.36) (red circles/line, right axis) and the modified McCarthy-Mason prediction (Equation 4.14) (green diamonds/line, right axis). A clear correlation between $\Gamma - 1$ and $\Delta D_{1,\mathrm{self,rel}}$ can be observed. The error bars of $\Gamma - 1$ are smaller than the symbol sizes. A full set of plots for all considered LJ systems is given in Appendices A.3.2-A.3.3.

highly non-ideal.

We can now compare the concentration dependencies of $(\Gamma - 1)$ and $\Delta D_{1,\mathrm{self,rel}}$. Figure 4.2 (bottom figure) suggests a strong correlation between $\Delta D_{1,\mathrm{self,rel}}$ and $(\Gamma-1)$: Large deviations of a mixture from ideal behavior lead to large relative deviations of the McCarty-Mason prediction.

To study the correlation between $\Delta D_{1,\mathrm{self,rel}}$ and $(\Gamma - 1)$ for the full set of LJ systems, we plot $\Delta D_{1,\mathrm{self,rel}}$ as function of $(\Gamma - 1)$ (cf. Figure 4.3(a)). A general trend $\Delta D_{1,\mathrm{self,rel}} \propto \Gamma - 1$ can be observed. However, the data scatters for molar mass ratios $m_2/m_1 > 2$. As a first approximation, we restrict our analysis to systems with molar mass ratios $m_2/m_1 < 2$. Figure 4.3(b) shows the relative deviation $\Delta D_{1,\mathrm{self,rel}}$ as function of $(\Gamma - 1)$ for all LJ systems with molar mass ratios $m_2/m_1 < 2$. A clear correlation $\Delta D_{1,\mathrm{self,rel}} \propto \Gamma - 1$ can be observed. For the full range of thermodynamic factors $0 < \Gamma < 7$, the McCarty-Mason predictions show large deviations of up to 130 %. The root-mean square error of $\Delta D_{i,\mathrm{self,rel}}$ is $RMSE = 35\,\%$, i.e. the McCarty-Mason predictions deviate by 35 % on average. However, molecular systems typically have thermodynamic factors in the range $0 < \Gamma < 2$ (cf. Section 4.2.1). Still, even in this molecular systems range $(0 < \Gamma < 2)$, the McCarty-Mason predictions have an $RMSE$ of 10 %.

To improve the McCarty-Mason predictions, we introduce a linear fit of $\Delta D_{i,\mathrm{self,rel}}$ as function of Γ following Equation 4.7. Fitting both $\Delta D_{1,\mathrm{self,rel}}$ and $\Delta D_{2,\mathrm{self,rel}}$ in the typical range of thermodynamic factors $0 < \Gamma < 2$ results in the function (cf. Figure 4.3(b))

$$f(\Gamma) = -0.2807 \cdot (\Gamma - 1)\,. \tag{4.13}$$

Insertion of Equation 4.13 into Equation 4.9 leads to an improved predictive equation for concentration-dependent self-diffusion coefficients:

$$\frac{1}{D_{i,\mathrm{self}}} = \left(\frac{x_1}{D_{i,\mathrm{self}}^{x_1 \to 1}} + \frac{x_2}{D_{i,\mathrm{self}}^{x_2 \to 1}} \right) \cdot \left(1 + 0.2807 \cdot (\Gamma - 1)\right). \tag{4.14}$$

In the following, Equation 4.14 will be called the "modified McCarty-Mason equation". Using the modified McCarty-Mason equation, the accuracy of the predictions is doubled compared to the McCarty-Mason predictions: The $RMSE$ halves from 10 % to 5 % for $0 < \Gamma < 2$.

The improved predictions of the modified McCarty-Mason prediction are visualized for the example LJ system considered in Figure 4.2: Using the modified version of the McCarty-Mason prediction decreases the maximum relative deviation from $\Delta D_{1,\text{self,rel}} = 70\,\%$ to $\Delta D_{1,\text{self,rel}} = 12\,\%$, i.e. the predictions are improved by a factor up to $0.7/0.12 = 5.8$. A full set of plots for all considered LJ systems is provided in Appendix A.3.

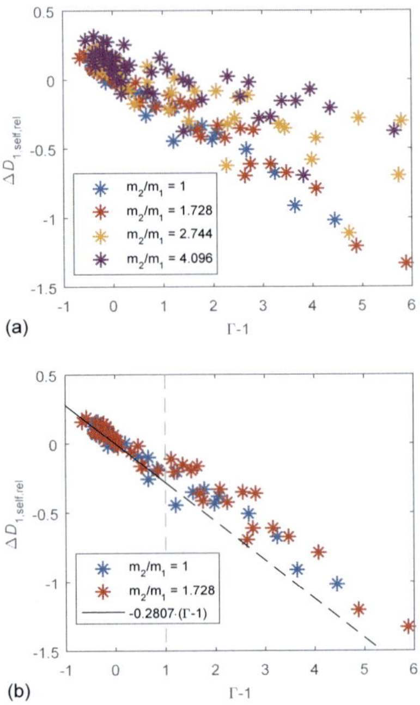

Figure 4.3: Relative deviations $\Delta D_{1,\text{self,rel}}$ of the McCarty-Mason predictions (Equation 2.36) of self-diffusion coefficients as function of the thermodynamic factor Γ for LJ systems.
(a) $\Delta D_{1,\text{self,rel}}$ for all LJ systems, color-coded by the molar mass ratios m_2/m_1.
(b) $\Delta D_{1,\text{self,rel}}$ for LJ systems with molar mass ratios $m_2/m_1 < 2$ and best fit of Equation 4.7 (black line) for $0 < \Gamma < 2$ (indicated by the vertical dashed line). Plots for the second species are provided in Appendix A.3.1.

Self-diffusion coefficients of molecular systems

The modified McCarty-Mason equation (Equation 4.14) was obtained from MD data of LJ systems. To evaluate the practical performance of the modified McCarty-Mason equation, it has to be tested with experimental data. Figure 4.4(a) shows the relative deviation $\Delta D_{1,\mathrm{self,rel}}$ of the McCarty-Mason equation (Equation 2.36) as function of $(\Gamma - 1)$. The correlation between $\Delta D_{1,\mathrm{self,rel}}$ and $(\Gamma - 1)$ is not as clear as for LJ systems (cf. Figure 4.3); even ideal mixtures $(\Gamma - 1 = 0)$ have relative deviations $\Delta D_{1,\mathrm{self,rel}} \neq 0$. Still, the linear fit Equation 4.13, which represents the predictions of the modified McCarty-Mason equation, captures a major part of the experimental data, but some molecular systems show large deviations. In particular, the systems water-N-methypyridine and methanol-carbon tetrachloride show large deviations with completely different dependencies of $\Delta D_{1,\mathrm{self,rel}}$ on $(\Gamma - 1)$. This plot suggests that it may be even impossible to derive a model based on the thermodynamic factor Γ only that can capture all molecular systems.

However, the modified McCarty-Mason equation was derived for systems with molar mass ratios $M_2/M_1 < 2$. In addition, it was shown in Section 4.2.1 that mixtures with dimerising species need a separate analysis. Excluding systems with $m_2/m_1 < 2$ and systems with dimerising species results in the remaining dataset shown in Figure 4.4(b). For the remaining dataset, a clear correlation between $\Delta D_{1,\mathrm{self,rel}}$ and $(\Gamma - 1)$ is observed, which agrees with the linear fit Equation 4.13 of the modified McCarty-Mason equation. The *RMSE* of the McCarty-Mason predictions is 11 %. If the modified McCarty-Mason equation is used, the *RMSE* decreases to 5 %. Hence, the deviations of the predictions made by the modified McCarty-Mason equation are $0.11/0.05 > 2$ times lower.

The improvement in the prediction of self-diffusion coefficients can also be visualized in terms of the concentration dependence: Figure 4.5 shows experimental and predicted concentration-dependent self-diffusion coefficients of the exemplary systems nitrobenze-hexane and cyclohexane-benzene (a full set of plots for all considered molecular systems is provided in Appendix A.4). For the system nitrobenze-hexane, the *RMSE* of the McCarty-Mason prediction is 23 % and the *RMSE* of the modified McCarty-Mason prediction is 6 %. Hence, the deviation of the predictions made by the modified McCarty-Mason equation are $0.23/0.06 \approx 4$ times lower. For the system cyclohexane-benzene, using the modified version of the McCarty-Mason equation de-

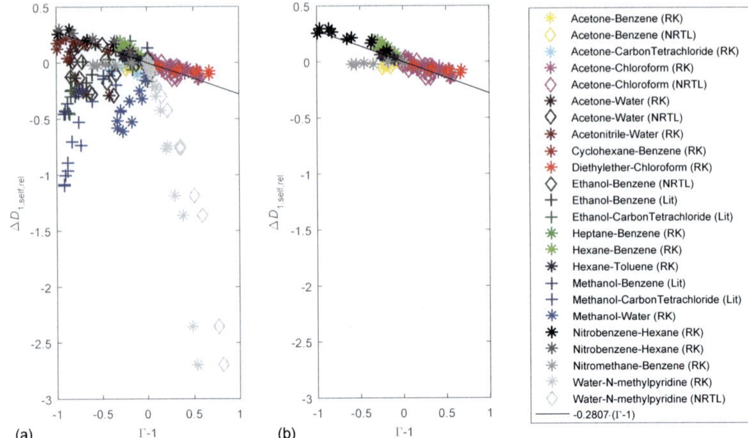

Figure 4.4: Relative deviations $\Delta D_{1,\text{self,rel}}$ of the McCarty-Mason prediction (Equation 2.36) as function of the thermodynamic factor Γ for molecular systems (symbols) and linear fit of $\Delta D_{1,\text{self,rel}}$ derived from LJ systems (black line, cf. Equation 4.13). Stars: Experimental data with thermodynamic factors calculated with Redlich-Kister (RK). Diamonds: Experimental data with thermodynamic factors calculated with NRTL. Plus symbols: Experimental data with thermodynamic factors reported in literature.
(a) $\Delta D_{1,\text{self,rel}}$ for all considered molecular systems.
(b) $\Delta D_{1,\text{self,rel}}$ for molecular systems with molar mass ratios $M_2/M_1 < 2$ and without dimerising species. Plots for the second species are provided in Appendix A.4.1.

creases the *RMSE* from 17 % to 4 %, which also corresponds to an improvement by a factor $0.17/0.04 \approx 4$.

Hence significant improvements in the prediction of concentration-dependent self-diffusion coefficients of non-ideal binary liquid mixtures are obtained by use of the modified McCarty-Mason equation. Combining the modified McCarty-Mason equation (Equation 2.36) with the Moggridge equation (Equation 2.33) and Equation 2.12 leads to the following model for the prediction of concentration-dependent binary Fick diffusion coefficients:

$$D_{12} = (x_2 D_{1,\text{self}} + x_1 D_{2,\text{self}})\,\Gamma^{0.64}, \tag{4.15}$$

$$\frac{1}{D_{i,\text{self}}} = \left(\frac{x_1}{D_{i,\text{self}}^{x_1\to 1}} + \frac{x_2}{D_{i,\text{self}}^{x_2\to 1}}\right) \cdot (1 + 0.2807 \cdot (\Gamma - 1))\,, \quad i = 1, 2. \tag{4.16}$$

Thus, to predict concentration-dependent Fick diffusion coefficients of binary mixtures with molar mass ratios $M_2/M_1 < 2$ and without dimerising species, we need only the self-diffusion coefficients at infinite dilution $D_{i,\text{self}}^{x_{j\neq i}\to 1} = D_{ij}^{x_{j\neq i}\to 1} = D_{12}^{\infty}$ and the self-diffusion coefficients of the pure substances $D_{i,\text{self}}^{x_i\to 1}$ as well as the thermodynamic factor Γ of the mixture.

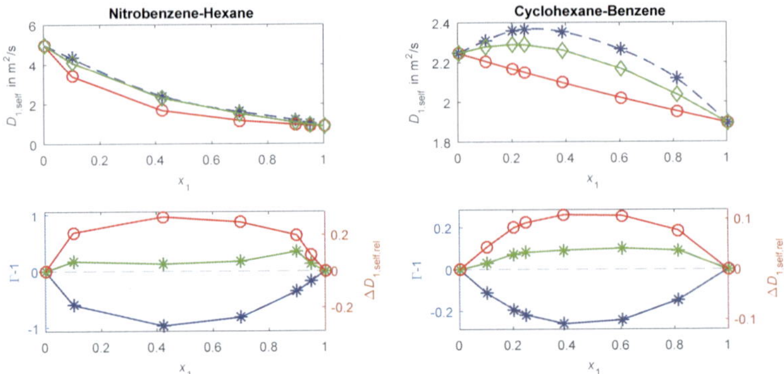

Figure 4.5: Concentration-dependent self-diffusion coefficients $D_{1,\text{self}}$, thermodynamic factors Γ, and relative deviations $\Delta D_{i,\text{self,rel}}$ for the systems nitrobenze-hexane (a) and cyclohexane-benzene (b). Note the adapted y-axis scale for each subfigure.
Top figures: Blue stars: Experimental data of concentration-dependent self-diffusion coefficients $D_{1,\text{self}}$. Blue-dashed line: smoothing fit of the experimental self-diffusion coefficients; red circles/line: predictions of the McCarty-Mason equation (Equation 2.36); green diamonds/line: predictions of the modified McCarty-Mason equation (Equation 4.14).
Bottom figures: Concentration dependence of the thermodynamic factor $(\Gamma - 1)$ (blue stars/line, left axis) and concentration dependence of the relative deviation $\Delta D_{i,\text{self,rel}}$ between the experimental self-diffusion coefficients and the McCarty-Mason predictions (Equation 2.36) (red circles/line, right axis) and the modified McCarthy-Mason prediction (Equation 4.14) (green diamonds/line, right axis).

4.3 Conclusions

The reliable prediction of concentration-dependent mutual diffusion coefficients has been a challenge to scientists for decades. For ideal mixtures, the physically-based Darken equation holds. For non-ideal mixtures, semi-empirical modifications of the Darken equation have been developed. However, Darken-based models rely on the knowledge of concentration-dependent self-diffusion coefficients which are rarely available.

Therefore, predictions of concentration-dependent self-diffusion coefficients are needed. In this chapter, we studied the concentration-dependence of mutual and self-diffusion coefficients in non-ideal binary liquid mixtures. The basis of our analysis were data of Lennard-Jones (LJ) systems from Molecular Dynamics simulations which provide insight into the full set of transport data and thermodynamic properties. For both mutual and self-diffusion, strong correlations between non-ideal diffusion effects and the thermodynamic factor were observed. The existing modification of the Darken equation by D'Agostino et al. (2011) and Moggridge (2012b) was confirmed to accurately predict concentration-dependent mutual diffusion coefficients for a wide range of non-ideal mixtures with typical thermodynamic factors ($0 < \Gamma < 2$). For mixtures with very large thermodynamic factors ($\Gamma > 2$), the data of the LJ systems suggest deviations.

Based on the predictive model of McCarty and Mason (1960) for ideal binary gas mixtures, we developed an improved model for the prediction of concentration-dependent self-diffusion coefficients in non-ideal binary liquid mixtures. Our new model is a function of the thermodynamic factor and the self-diffusion coefficients at infinite dilution and of the pure substances which are readily available. Validation was carried out with experimental data of molecular systems. Self-diffusion coefficients of mixtures with typical thermodynamic factors $\Gamma < 2$, molar mass ratios $m_2/m_1 < 2$, and without dimerising species are successfully predicted: The relative deviation of the predictions is halved from 10 % to 5 %. Our new model thus provides the missing link to render Darken-based models into practical tools to predict mutual diffusion coefficients.

Chapter 5

Model-based optimal experimental design of diffusion experiments

Reprinted (adapted) with permission from Elsevier from
Wolff, L., Koß,H.-J., and Bardow, A. (2016). The optimal diffusion experiment, *Chem. Eng. Sci.*, 152:392-402. (Wolff et al., 2016)
Contribution of the author: Writing the draft, principal author, modeling, simulation, analysis.

Even with improved predictive models for diffusion coefficients as introduced in the preceding chapter, the scarcity of diffusion data cannot be entirely bypassed: Predictive models are still limited in accuracy and limited to certain mixtures, or temperature and pressure contitions. In addition, predictive models need validation. Therefore, experiments are indispensable and still the most significant source for diffusion coefficients. However, even by use of the most advanced equipment, diffusion experiments are time-consuming and expensive (cf. Chapter 2). Thus, a reduction of the *number* of required diffusion experiments would be desirable. At the same time, the accuracy of the diffusion coefficients should be retained or even enhanced. This goal can be achieved by use of the method of model-based optimal experimental design (OED), which identifies the experiments with the highest information content (cf. Chapter 3). OED has already been successfully applied to diffusion experiments in diffusion cells and microfluidic channels (Bardow et al., 2003, 2005a, 2006; Kriesten et al., 2009; Häusler et al., 2012). Thereby, optimal measurement times and measurement positions have been identified and verified for diffusion experiments. These successfull applications of OED for diffusion lead to a fundamental question: Which diffusion experiment provides the most information on diffusion coefficients?

In general, diffusion experiments can be characterized by two properties: Analytical technique and diffusion geometry. Analytical techniques and their accuracy are subject to continuous improvement. The accuracy of analytical techniques is taken into account in OED. Thus, OED could be employed to assess and rank current analytical techniques. However, this ranking would only represent the current state of the art. No general conclusion can be drawn which analytical technique will be optimal tomorrow and in the longer future. In contrast, the choice of geometry can be investigated independently from the analytical technique on a purely theoretical basis. Even geometries for which no realization exists today can be investigated. Therefore, we can identify the geometries which offer the optimal conditions for the precise determination of diffusion coefficients.

In this chapter, we identify optimal and robust diffusion geometries for binary diffusion measurements. A diffusion geometry is defined by its shape and boundary conditions for the diffusion process. Considered shapes are planes, cylinders, and spheres. As boundary conditions, walls with zero flux conditions and surfaces with constant concentration are considered. The influence of shapes and boundary conditions on the efficiency of diffusion measurements is analyzed using analytical equations and numerical computations. We further study the potential impact of uncertainties in the specification of the experiment on the accuracy of diffusion coefficients to identify robust designs.

In Section 5.1, the diffusion model, the considered diffusion geometries, and the OED method for diffusion measurements are introduced. The results of the OED are given in Section 5.2: In Subsection 5.2.1, we study the influence of different boundary conditions on the efficiency of diffusion experiments. In Subsection 5.2.2, quantitative results for the optimized experiments are given. Moreover, in Subsection 5.2.3, we study the influence of errors in the independent variables - i.e., measurement positions, times, and initial domain boundaries - on the error in diffusion coefficients. Thereby, robust geometries are identified which are insensitive to errors in the independent variables.

5.1 Method

The aim of OED is to identify an experimental design which minimizes the uncertainties - in statistical terms: the (co-)variances - of the model parameters, i.e., here the diffusion coefficients (cf. 3). In this chapter, we investigate diffusion in binary mixtures with ideal mixing behavior and a single concentration-independent diffusion coefficient D = const. These assumptions are typically fulfilled in diffusion experiments by studying diffusion across a small change in concentration $\Delta c = c_0 - c_1$ (Woolf et al., 1991). Binary diffusion represents the base case and is most frequently studied. In our experience, findings related to geometry can be transferred from binary to multicomponent diffusion (Bardow et al., 2006).

In the scope of this chapter, a diffusion experiment consists of the measurement of concentrations at defined positions and instants of time for a given geometry. A variety of diffusion geometries as well as spatial and temporal resolutions are taken into account. Since our focus is on diffusion geometries, we identify optimal measurement positions and times for each diffusion geometry. Subsequently, we compare the optima of all diffusion geometries to classify the performance of the diffusion geometries. Thereby, optimal diffusion geometries are identified. Since OED is a model-based procedure, we need a model for each geometry to relate the concentrations to position and time.

5.1.1 Geometries for diffusion experiments

In the scope of this chapter, we investigate one-dimensional diffusion processes in which mass transport occurs only due to Fick diffusion, i.e., no convective mass transport and no excess volume effects are present such that the volume average velocity is zero. The differential mass balance for diffusion in a plane in the volumetric reference frame for component 1 of a binary mixture reads as follows (cf. Equation 2.17):

$$\frac{\partial c(z,t)}{\partial t} = D\frac{\partial^2 c(z,t)}{\partial z^2}. \tag{5.1}$$

To generalize the investigation, a dimensionless form of the mass balance Equation (5.1) is used:

$$\frac{\partial \bar{c}(\xi, Fo)}{\partial Fo} = \frac{\partial^2 \bar{c}(\xi, Fo)}{\partial \xi^2}. \tag{5.2}$$

Here, $\bar{c} = (c - c_1) / (c_0 - c_1) \in [0, 1]$ is the concentration difference $c - c_1$ scaled by the maximum concentration difference $c_0 - c_1$ in the beginning of the diffusion process. $\xi = z/L \in [0, 1]$ is the position z scaled by the total length L, and $Fo = (Dt)/L^2 \in [0, \infty)$ is the Fourier number, which represents a dimensionless time. For cylinders and spheres, similar differential mass balances hold by inserting the appropriate coordinate system (see Crank (1975)).

The solutions $\bar{c}(\xi, Fo)$ of the differential mass balances as a function of position ξ and time Fo depend on the shape of the diffusion geometry G as well as on the initial and boundary conditions. We consider shapes of elementary mathematical forms: plane, cylinder, and sphere. The initial conditions are defined by discrete concentration domains in the beginning of the diffusion process as shown in Figure 2.1(a). The position of the initial domain boundary is denoted $\xi_0 = z_0/L$. Two boundary conditions are studied: a zero flux condition, $\partial \bar{c}/\partial \xi = 0$, and a zero concentration condition, $\bar{c} = 0$. The zero flux condition represents a wall or a symmetry axis. The zero concentration condition represents a surface with constant concentration $c = c_1$.

With the above-mentioned shapes and boundary conditions, two categories of geometries can be introduced:

1. Geometries with two walls perpendicular to the diffusion direction, i.e., where both boundary conditions are zero flux conditions (cf. Figure 5.1, top row):
   a) The free diffusion geometry (cf. Figure 5.1(a)) is the limiting case for diffusion in a plane with two walls. Two semi-infinite diffusion domains are put into contact at the position ξ_0 of the initial domain boundary. The walls are infinitely far away.
   b) When the walls are placed at a finite distance from the position of the initial domain boundary, a closed diffusion cell is obtained (Figure 5.1(b)). In the same manner, the closed cylinder (Figure 5.1(d)) and the closed sphere (Figure 5.1(e)) are obtained. These closed geometries have zero flux boundary conditions at the walls and symmetry axes, $\xi = 0$ and $\xi = 1$. The initial domain boundary is inside the geometry, all mass transport takes places inside the geometry. The microfluidic channel (Figure 5.1(c)) is

similar to the closed cell: Here, the Fourier number represents the average residence time in the channel which is a function of the position ξ due to the velocity profile inside the channel.

2. Geometries with one wall, i.e., one zero flux boundary condition, and one surface with constant concentration as boundary condition (cf. Figure 5.1, bottom row):
   a) The semi-free diffusion geometry (cf. Figure 5.1(f)) is the limiting case for diffusion in a plane with one wall and one surface with constant concentration. Here, a semi-infinite diffusion domain has a wall at $\xi \to -\infty$ and a surface with constant concentration at $\xi = 1$.
   b) When the wall is placed at a finite distance from the surface with constant concentration, open geomeries are obtained (cf. Figures 5.1(g-i)). Again, diffusion in a cell, a cylinder, and a sphere is considered. The wall is at $\xi = 0$, the surface with constant concentration at $\xi = 1$.

The concentration profiles $\bar{c} = f(G, Fo, \xi, \xi_0, D)$ for all geometries G are given and illustrated in Appendices B.1 and B.2. Furthermore, solutions for the time-dependent average concentrations $\bar{c}_{\text{av}} = \bar{c}_{\text{av}}(G, Fo)$ of the open geometries are given which represent the spatial average obtained by integrating the concentrations over the domain. These average concentrations are measured in some established diffusion experiments, e.g. the open capillary method (Anderson and Saddington, 1949) (cf. Section 2.1).

The generic representation of the diffusion experiments allows to study practically all geometries already employed or conceivable. Noteworthy exceptions are

1. the diaphragm cell which studies a quasi-steady state concentration gradient and requires calibration and leads to excessively long experiments (Woolf et al., 1991); and
2. dynamic light scattering (DLS) which studies diffusion in a homogeneously mixed system (Leipertz and Fröba, 2005) but only yields the eigenvalues of the Fick matrix in multicomponent mixtures (Bardow, 2007; Heller et al., 2016).

The systematic overview of geometries includes both geometries already employed for diffusion experiments and novel geometries. A literature overview is given in Appendix B.3 showing examples where different geometries have already been employed in practice.

We do not consider the trade-off between experimental effort for setting up geometries and the information content of experiments conducted in these geometries. A method to deal with this kind of trade-off is descibed by Telen et al. (2013).

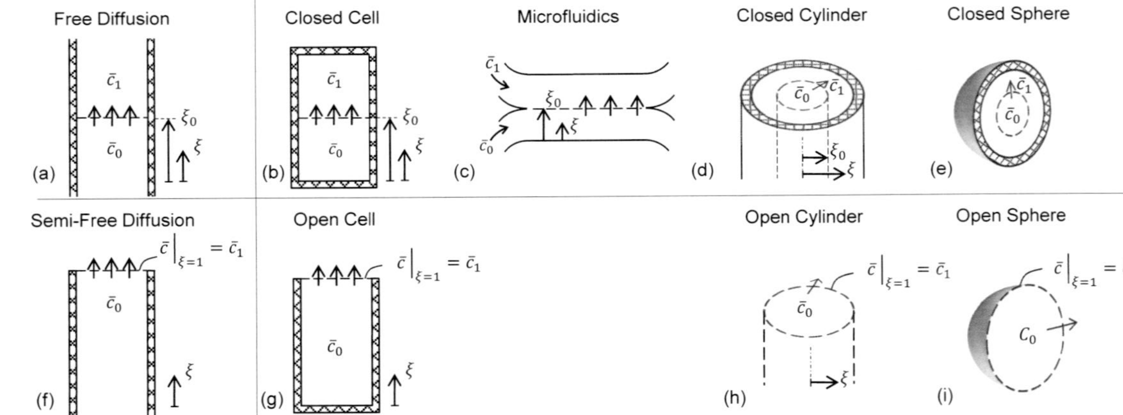

Figure 5.1: Diffusion geometries with initial conditions. Hatched areas illustrate walls; dashed lines illustrate the position of the initial domain boundary. Upper row: Diffusion geometries where both boundary conditions are zero flux conditions. Lower row: Diffusion geometries where one boundary condition is a zero flux condition, and the other boundary condition is a surface with constant concentration.

5.1.2 Model-based optimal experimental design

The mathematical solutions $\bar{c} = f(G, Fo, \xi, \xi_0, D)$ are the basis for the optimal experimental design analysis (cf. Section 3). The aim of our analysis is to identify an experimental design which minimizes the uncertainty in the model parameter, i.e., the diffusion coefficient D. The experimental design $\boldsymbol{d}$ is defined by the independent variables, i.e., the geometry G, the measurement time Fo, the measurement position ξ, and the position ξ_0 of the initial domain boundary (ξ_0 only if closed geometries are considered):

$$\boldsymbol{d} = \boldsymbol{d}(G, Fo, \xi, \xi_0). \tag{5.3}$$

Regarding the properties of the model and the measurement errors in the concentrations $\bar{c}$, we make the assumptions introduced in Section 3.1:

1. The model is of exact-structural type. This assumption seems to be well justified based on the fact that Fick's 2nd law (Equation 5.1) has been the basis for different experimental techniques leading to consistent diffusion data (see reviews by Peters et al. (2016); Woolf et al. (1991)).

2. The model parameter (i.e., the diffusion coefficient) is identifiable, i.e., the diffusion coefficient can be identified from the measurements. Identifiability of diffusion coefficients in models of the form present in this work was proven by Bardow (2004).

3. There is no systematic error in the measurements.

4. The measurement errors in different experiments are independent and normally distributed with the variance $\sigma_{\bar{c}}$.

For measurements at one instant of time and one position, the Fisher information F is equal to the local Fisher information M (cf. Equations 3.8 and 3.9):

$$F(\boldsymbol{d}, D) = M(G, Fo, \xi, \xi_0, D) = \left(\frac{\partial \bar{c}}{\partial D}\right)^2 \cdot \frac{1}{\sigma_{\bar{c}}^2}. \tag{5.4}$$

The local Fisher information M is high if the model output $\bar{c}$ is sensitive to changes in the model parameter D, and if the measurement uncertainty $\sigma_{\bar{c}}^2$ is low. The derivative

$\partial \bar{c}/\partial D$ is calculated using the chain rule:

$$\frac{\partial \bar{C}}{\partial D} = \frac{\partial \bar{C}}{\partial Fo} \cdot \frac{\partial Fo}{\partial D} = \frac{\partial \bar{C}}{\partial Fo} \cdot \frac{Fo}{D} \tag{5.5}$$

The measurement variance $\sigma_{\bar{c}}^2$ is assumed to be constant. Thereby, we could account explicitly for the accuracy of different analytical techniques. However, we do not study the impact of different analytical techniques in this work because the accuracies are continuously improved. Thus, the conclusions would only represent a snapshot in time. In addition, the general conclusion is trivial: The smaller the measurement variance $\sigma_{\bar{c}}^2$, the higher the Fisher information. Thus, the analytical technique with the highest accuracy would be preferred. We would only expect trade-offs if analytical techniques would show different variances for different diffusion geometries or if analytical techniques would show inhomogeneous variances depending on the experimental design. In these cases, an inhomogeneous variance $\sigma_{\bar{c}}^2(\boldsymbol{d})$ could also be included in the design formulation (for details consult the book of Bard (1974)). A potential trade-off could arise, e.g., if the most accurate analytical techniques would be limited to non-optimal diffusion geometries or if the measurement accuracy deteriorates for design settings leading to large sensitivities. However, the trade-offs would be very case specific and therefore beyond the scope of the present study.

Till here, only measurements at one position and one instant of time have been considered. However, different spatial and temporal resolutions are possible. Considering the spatial resolution, a localized measurement at one position (e.g., Lin et al. (2010)), a continuous measurement along a line (e.g., Bardow et al. (2003)), or a non-spatially resolved measurement of average concentrations is possible (e.g. (Anderson and Saddington, 1949). Considering the temporal resolution, a measurement at one time, or a continuous measurement over time is possible. Therefore, corresponding formulations are introduced for the Fisher information (Equation 5.4) and the design vector (Equation 5.3). For continuous measurements, an integral measure of the Fisher information is used, following a methodology introduced by Galvanin et al. (2011). By combining the different spatial and temporal resolutions, the following 5 additional formulations for the Fisher information are obtained:

1. Measurement of concentration at one position ξ and continuously over time

$Fo = [0, \infty)$ ($\boldsymbol{d} = \boldsymbol{d}(G, \xi, \xi_0)$):

$$F(G, \xi, \xi_0, D) = \int_0^\infty M(G, Fo, \xi, \xi_0, D)\,\mathrm{d}Fo, \tag{5.6}$$

2. Spatially-resolved measurement of concentration over $\xi = [0, 1]$ at one instant of time Fo ($\boldsymbol{d} = \boldsymbol{d}(G, Fo, \xi_0)$):

$$F(G, Fo, \xi_0, D) = \int_0^1 M(G, Fo, \xi, \xi_0, D)\,\mathrm{d}\xi, \tag{5.7}$$

3. Spatially-resolved measurement of concentration over $\xi = [0, 1]$ continuously over time $Fo = [0, \infty)$ ($\boldsymbol{d} = \boldsymbol{d}(G, \xi_0)$):

$$F(G, \xi_0, D) = \int_0^\infty \int_0^1 M(G, Fo, \xi, \xi_0, D)\,\mathrm{d}\xi\,\mathrm{d}Fo, \tag{5.8}$$

4. Average concentration measurements averaged over $\xi = [0, 1]$ at one instant of time Fo (for open geometries only, $\boldsymbol{d} = \boldsymbol{d}(G, Fo)$):

$$F(G, Fo, D) = M_{\mathrm{av}}(G, Fo, D), \tag{5.9}$$

5. Average concentration measurements averaged over $\xi = [0, 1]$ continuously over time $Fo = [0, \infty)$ (for open geometries only, $\boldsymbol{d} = \boldsymbol{d}(G)$):

$$F(G, D) = \int_0^\infty M_{\mathrm{av}}(G, Fo, D)\,\mathrm{d}Fo. \tag{5.10}$$

The position ξ_0 of the initial domain boundary influences closed geometries only. Therefore, it drops for the objective functions Equations 5.6-5.10, if open geometries are considered.

Knowledge of the Fisher informations enables the calculation of the optimal design and the resulting D-efficiencies ζ (Equation 3.23) for the different geometries as well as for different spatial and temporal resolutions. Note that the D-efficiency is independent of the value of the diffusion coefficient D: The local Fisher information M (and thus the Fisher information F) is proportional to $(1/D)^2$ (Equations 5.4-5.5). Thus, the dependence on D cancels in the calculation of the D-efficiency which is a ratio of two Fisher information scalars F (Equation 3.23).

5.2 Results and discussion

In this section, we discuss the D-efficiencies ζ and optimal designs for the different geometries for different spatial and temporal resolutions. In Section 5.2.1, we conduct a qualitative analysis of the D-efficiencies for measurements at one position and one instant of time for all geometries. The influence of shapes and boundary conditions on the D-efficiencies is analyzed. In Section 5.2.2, we conduct a quantitative analysis by computing optimal experimental designs for all objective functions defined in Section 5.1.2. Thereby, optimal measurement positions, optimal measurement times, and optimal positions of the initial domain boundary are identified for each geometry and each objective function. A subsequent comparison of the D-efficiencies enables the identification of the optimal geometry for each objective function. In Section 5.2.3, the influence of errors in the independent variables on the D-efficiencies is discussed. Independent variables are the variables which are manipulated by the experimenter, i.e., measurement positions, measurement times, and positions of the initial domain boundary. We identify robust optimal geometries, which are insensitive to errors in the independent variables and enable precise estimations of diffusion coefficients.

5.2.1 Impact of geometry and boundary conditions

In this section, the influence of geometry shapes and boundary conditions on the D-efficiency of diffusion experiments is analyzed and general design rules for diffusion experiments are derived. The geometries of Figure 5.1 are discussed from left to right and from top to bottom. Thereby, the influence of walls, radial distortions, and constant concentrations at the geometry surface are analyzed.

Base-case: Free diffusion

Free diffusion experiments are most commonly employed (Woolf et al., 1991; Cussler, 2007). Free diffusion allows to facilitate the analysis of experimental results, because a simple analytical solution of the concentration profile is available. The solution for the dimensionless concentration profile is (Carslaw and Jaeger, 1959):

$$\bar{c} = \frac{1}{2}\left[1 - \mathrm{erf}\left\{\frac{\xi - \xi_0}{2\sqrt{Fo}}\right\}\right]. \tag{5.11}$$

A graphical illustration of the concentration profile for $\xi_0 = 0.5$ is provided in Appendix B.2. By application of Equation 5.4, the Fisher information for measurements at one position and one instant of time can be calculated:

$$F = \frac{1}{\sigma_{\bar{c}}^2} \cdot \frac{(\xi - \xi_0)^2}{16\pi D^2 Fo}\left[\exp\left\{-\left(\frac{\xi - \xi_0}{2\sqrt{Fo}}\right)^2\right\}\right]^2. \tag{5.12}$$

The D-efficiency as a function of position ξ and time Fo is shown in Figure 5.2(a). It has a maximum along a ridge, which can be described by

$$Fo = \frac{(\xi - \xi_0)^2}{2}. \tag{5.13}$$

Beyond this ridge, the D-efficiency decreases considerably.

Closed cell: Influence of walls

Free diffusion is the limiting case for diffusion in a cell as long as the walls have no or little influence on the diffusion process, i.e., for short times or most precisely small Fourier numbers Fo. The influence of walls (i.e., boundary conditions $\partial\bar{c}/\partial\xi = 0$) is revealed by comparison of the case of free diffusion with diffusion in a closed cell. The analytical solution for the dimensionless concentration profile in the closed cell with walls at the positions $\xi = 0$ and $\xi = 1$ is given by (Carslaw and Jaeger, 1959)

$$\bar{C} = \frac{1}{2} \cdot \sum_{n=-\infty}^{\infty}\left(\mathrm{erf}\left\{\frac{\xi_0 + 2n - \xi}{2\sqrt{Fo}}\right\} + \mathrm{erf}\left\{\frac{\xi_0 - 2n + \xi}{2\sqrt{Fo}}\right\}\right). \tag{5.14}$$

Equation 5.14 can be derived from the free diffusion case (Equation 5.11) by reflecting the concentration profile at the walls (Crank, 1975). Due to the walls, the concentration equilibration is much faster than in the case of free diffusion (cf. Appendix B.2).

The D-efficiencies of the closed cell are shown in Figure 5.2(b). For small diffusion times ($Fo < 0.02$), the D-efficiency shows the same ridge as in the case of free diffusion since the impact of the walls is not yet apparent and the diffusion process is practically free. For longer diffusion times ($0.02 < Fo < 0.3$), the reflection at the walls leads to an amplification of the D-efficiency at the walls, where the maximum at $Fo = 0.1$ is about 4 times larger than in the case of free diffusion. Therefore, measurements at the wall are highly advantageous for precise diffusion measurements. This result has already been reported and experimentally validated in a previous work for a 1-dimensional cell (Bardow et al., 2003). The D-efficiencies of the microfluidic channel are practically identical to the D-efficiencies of the closed cell (cf. Figure 5.2(c)). This result has also been reported and experimentally validated in a previous work of Häusler et al. (2012).

Influence of radial distortion

In addition to the influence of walls, the closed cylinder and the closed sphere introduce radial distortion to the diffusion process. The closed cylinder can be interpreted as a closed cell where, on the one hand, the wall of the closed cell at $\xi = 0$ is compressed into a line which becomes the symmetry axis of the cylinder, and on the other hand, the wall at $\xi = 1$ is wound around this symmetry axis and becomes the exterior wall of the cylinder. Hence, the diffusion area increases in radial direction from the symmetry axis to the external wall of the cylinder. The diffusion process is again faster than for the closed cell (cf. Appendix B.2). A change in the diffusion coefficient has a large impact on the change in the concentration at the symmetry axis ($\partial \bar{c}/\partial D$ is large). Therefore, the D-efficiency at the symmetry axis $\xi = 0$ is increased in comparison to the closed cell; the maximum is more pronounced and sharpened (see Figure 5.2(d)).

This effect is even stronger for diffusion in a closed sphere: The sphere can be interpreted as a cylinder, whose symmetry axis $\xi = 0$ is compressed into a point, the symmetry point of the sphere, and whose exterior wall becomes the exterior surface of the sphere. The diffusion process is fastened again (see Appendix B.2) and the maximum D-efficiency at $\xi = 0$ is even more amplified and sharpened (cf. Figure 5.2(e)).

Altogether, radial distortion of a geometry leads to increased and sharpened D-efficiencies ζ at the symmetry axis.

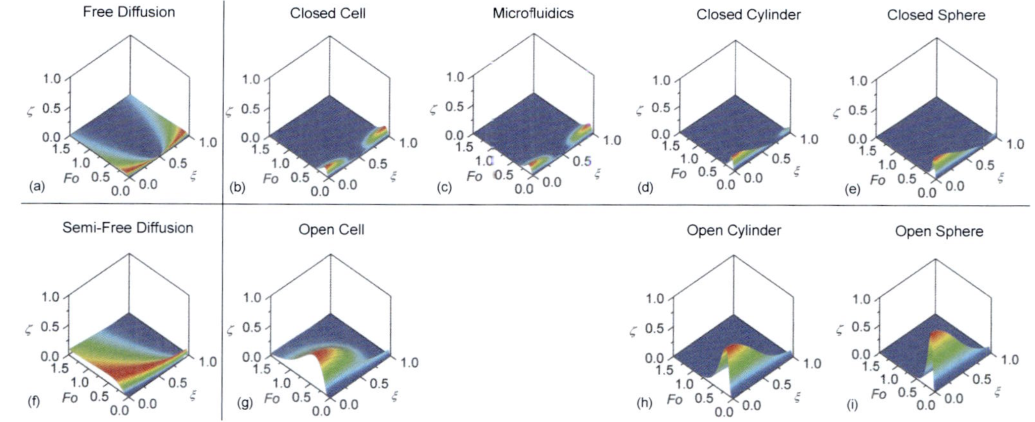

Figure 5.2: Information in diffusion experiments as measured by the D-efficiencies ζ_D (Equation 3.23) for all considered geometries (Figure 5.1) as function of dimensionless position ξ and dimensionless time (represented by the Fourier number Fo). To allow for a clear representation, the positions ξ_0 of the initial domain boundary of the free diffusion geometry and the closed geometries (upper row) are not optimized here but chosen such that equal amounts of material are on both sides of the initial domain boundary.

Influence of constant surface concentrations

To investigate the influence of constant concentrations at the geometry surface, semi-free diffusion is compared to free diffusion. In contrast to the case of free diffusion, the semi-free diffusion geometry has a constant concentration $\bar{c} = 0$ at $\xi = 1$. The analytical solution for the concentration profile of the semi-free diffusion geometry is (Carslaw and Jaeger, 1959):

$$\bar{c} = -\text{erf}\left\{\frac{\xi - 1}{2\sqrt{Fo}}\right\}. \tag{5.15}$$

In comparison to the solution for the free diffusion case (Equation 5.11), Equation 5.15 shows a factor 2 in the concentration profile formula, leading to a factor 4 in the maximum D-efficiency (cf. Figure 5.2(f)).

Thus, one optimal diffusion experiment in a semi-free geometry would yield the same accuracy as 4 optimal experiments using free diffusion conditions. In case of the open cell, the additional positive influence of the wall on the D-efficiency can be observed just as for the closed geometries (cf. Figure 5.2(g)). The maximum D-efficiency is even more amplified for open cylinder (cf. Figure 5.2(h)) and open sphere (cf. Figure 5.2(i)) due to the additional influence of the radial distortion.

In summary, the maximum D-efficiency ζ_{D} of all geometries can be observed for the open sphere geometry, where the influences of wall, constant surface concentration, and maximum radial distortion are combined.

Measurement of average concentrations

Open geometries also allow for the measurement of average concentrations. For average concentration measurements, the D-efficiencies are a function of time. The D-efficiencies for the open cell, open cylinder, and open sphere are shown in Figure 5.3. In contrast to D-efficiencies of local concentration measurements, the radial distortion has a negative impact on the D-efficiencies of average concentration measurements. Therefore, the maximum D-efficiency is largest for the open cell and decreases from open cylinder to open sphere.

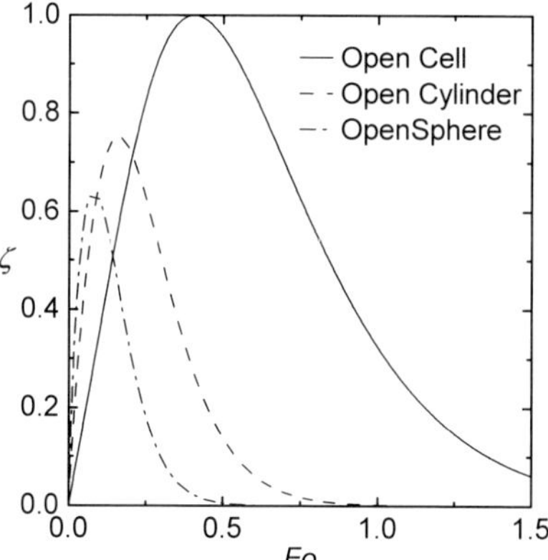

Figure 5.3: Information in diffusion experiments as measured by the D-efficiencies ζ_D (Equation 3.23) for average concentration measurements in open geometries as function of dimensionless time represented by the Fourier number Fo.

5.2.2 Optimized experiments

In this section, the optimal experimental designs are discussed for all geometries with respect to all spatial and temporal resolutions defined in Section 5.1.2. The optimization yields maximum D-efficiencies for each geometry and spatial and temporal resolution. The maximum D-efficiencies are shown in Figure 5.4. The corresponding values of the maximum D-efficiencies as well as the optimal measurement positions, measurement times, and positions of the initial domain boundaries are given in Appendix B.4.

Free diffusion and semi-free diffusion equations are valid only as long as the walls have no or little influence on the diffusion process. The exact solution, considering the influence of the walls, is given by the equations of the closed cell and the open cell, respectively. In the optimization, a maximum error of 1% between concentrations in the free diffusion/semi-free diffusion geometry in comparison to the closed cell/open cell was allowed.

Due to different objective functions, only results within one column of Figure 5.4 can be compared.

For measurements at one position and at one time, the open sphere geometry is identified as the most beneficial diffusion geometry (cf. Section 5.2.1). Therefore, its maximum D-efficiency is $\zeta_{\mathrm{D}} = 1$. In contrast, the free diffusion geometry has a maximum D-efficiency of $\zeta_{\mathrm{D}} = 0.0345$ only (cf. Appendix B.4). Thus, $1/0.0345 = 29$ times more measurements are required in the free diffusion geometry to obtain the same accuracy in the diffusion coefficient as from 1 measurement in the open sphere geometry. This analysis already assured optimal measurement positions, measurement times, and initial domain boundary positions in both geometries. The result may also be interpreted from the view of expected parameter uncertainty: If a diffusion coefficient of $D = 10^{-9}\,\mathrm{m}^2/\mathrm{s}$ is to be measured with a measurement uncertainty in the concentrations of $\sigma_{\tilde{c}} = 2\,\%$, the maximum Fisher information of the open sphere is $F_{\mathrm{opt}} = 1.06 \cdot 10^{21}\ (\mathrm{m}^2/\mathrm{s})^{-2}$. Therefore, the expected relative uncertainty in the diffusion coefficient is (see Equation 3.7)

$$\frac{\sigma_D}{D} = \frac{1/\sqrt{F_{\mathrm{opt}}}}{D} = 3.1\,\%. \tag{5.16}$$

When a free diffusion experiment is used instead, a relative uncertainty in the diffusion coefficient of

$$\frac{\sigma_D}{D} = \frac{1/\sqrt{\zeta_{\mathrm{D}} \cdot F_{\mathrm{opt}}}}{D} = 16.4\,\% \tag{5.17}$$

would have to be expected.

In general, the results show that open geometries have the highest information contents (cf. Figure 5.4). Employing open geometries instead of free diffusion experiments can improve the accuracy of the Fick diffusion coefficients by up to two orders of magnitude: The D-efficiencies of free diffusion experiments are are of the order 0.01 (cf. Appendix B.4). For measurements over time, geometries with no radial distortion, i.e., open cells, are advantageous. For measurements at one time, geometries with radial distortions, i.e., open spheres, are advantageous. If only closed geometries are considered, the closed sphere geometry is always the most beneficial one. For measurements of average concentrations, the open cell geometry is superior. In all cases, measurements at the walls provide the maximum information (cf. Appendix B.4).

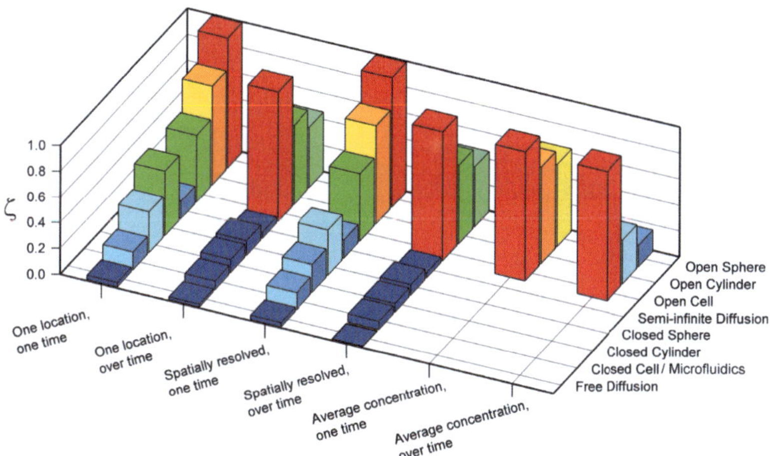

Figure 5.4: Information in diffusion experiments as measured by the D-efficiencies ζ_D (Equation 3.23) for the optimal experiments for each geometry and all spatial and temporal resolutions. Due to different objective functions, only results within one column can be compared to each other. Detailed values of the D-efficiencies are listed in Appendix B.4.

5.2.3 Errors in the independent variables Fo, ξ, ξ_0

In real applications, not only errors in the concentration measurement $\bar{c}$ have to be expected, but errors in the settings of the independent variables, specified by the experimenter, have to be expected as well. By incorporating these errors in the design formulation, we can identify robust optimal designs which are insensitive to errors in the independent variables. The independent variables specified by the experimenter are the measurement time, represented by the Fourier number Fo, the measurement position ξ, and the position ξ_0 of the initial domain boundary.

In Appendix B.5, a detailed scenario for errors in the independent variables is presented. In summary, the following characteristics are observed:

- For closed geometries, errors in the position ξ_0 of the initial domain boundary are the main cause for efficiency losses.
- For open geometries, errors in the Fourier number Fo are the main cause for

efficiency losses because ξ_0 is no variable in these geometries.

- Both error contributions in ξ_0 and Fo increase from free/semi-free diffusion over cell and cylinder to the sphere. Thus, this decrease in efficiency due to errors in the independent variables runs in the opposite direction as the increase of efficiency due to walls and radial distortion (cf. Section 5.2.1). Hence, there is a trade-off between both effects; the specific result depends on the magnitude of errors in the independent variables.
- Errors in the measurement position ξ are of minor importance for optimal measurements, because the walls are always the optimal measurement position.
- The free diffusion ridge for small Fo-numbers disappears for all geometries. Hence, free diffusion experiments are particularly severely affected by errors in the independent variables.

The results show that open geometries are still the most beneficial geometries in general. The open cell shows the highest efficiencies in most cases. However, the efficiencies of the open sphere, which was the former optimal geometry, are severely affected by errors in the independent variables. Hence, the open cell is the most robust diffusion geometry.

5.3 Conclusions

In this chapter, optimal geometries for one-dimensional binary diffusion measurements are identified using model-based optimal experimental design (OED) methods. Already existing as well as theoretically conceivable geometries are investigated; diffusion in planes, cylinders, and spheres is considered. We analyze the influence of walls (or symmetry axes, respectively) and surfaces with constant concentrations as boundary conditions. Thereby, two categories of geometries are defined: closed geometries with two walls (or one wall and one symmetry axis, respectively), and open geometries with one wall and one surface with constant concentration.

The open sphere with a constant surface concentration is identified as the most beneficial geometry for measurements at one time during the diffusion process. The only exception are measurements of average concentrations: For measurements of average concentrations as well as for continuous measurements over time, the open cell, featuring a wall and a surface with constant concentration, is best suited for diffusion measurements. Both optimal geometries belong to the class of open geometries.

Errors in the independent variables, specified by the experimenter, can affect the results. Considered independent variables are measurement time, measurement position, and the position of the initial domain boundary. Even in case of errors in the independent variables, the class of open geometries remains the most beneficial. The open cell is found to be the most robust geometry with regard to errors in the independent variables.

In general, a combination of walls and surfaces with constant concentration is identified as advantageous. Radial distortion, present in the cylindrical and spherical geometries, is disadvantageous for continuous measurements over time, but can be highly advantageous for one-time measurements. When optimal geometries are used instead of typically performed free diffusion experiments, the accuracy of the Fick diffusion coefficients can be improved by up to two orders of magnitude. Choosing the right geometry can therefore contribute significantly to the efficiency of diffusion measurements.

CHAPTER 6

Experimental analysis of diffusion experiments

Major parts of this chapter have been published in
Wolff, L., Zangi, P., Brands, T., Rausch, M. H., Koß, H.-J., Fröba, A. P., and Bardow, A. (2018). Concentration-dependent diffusion coefficients of binary gas mixtures using a Loschmidt cell with holographic interferometry, Part I: Multiple experiments, *Int. J. Thermophys.*, 39:133. Reproduced with permission of Springer Nature. (Wolff et al., 2018b)
Contribution of the author: Writing the draft, principal author, analysis of the interference data derived from diffusion experiments conducted by P. Zangi, model development and parameter estimation.

Wolff, L., Zangi, P., Brands, T., Rausch, M. H., Koß, H.-J., Fröba, A. P., and Bardow, A. (2018). Concentration-dependent diffusion coefficients of binary gas mixtures using a Loschmidt cell with holographic interferometry, Part II: Single experiment, *Int. J. Thermophys.*, 39:132. Reproduced with permission of Springer Nature. (Wolff et al., 2018c)
Contribution of the author: Writing the draft, principal author, analysis of the interference data derived from diffusion experiments conducted by P. Zangi, model development and parameter estimation.

Peters, C., Wolff, L., Haase, S., Thien, J., Brands, Koß, and Bardow, A. (2017). Multicomponent diffusion coefficients from microfluidics using Raman microspectroscopy, *Lab Chip*, 16:2768-2776. Reproduced with permission of The Royal Society of Chem-

istry. (Peters et al., 2017)
Contribution of the author: Co-author in writing the draft and final text, development of the diffusion model for multicomponent diffusion in a microfluidic H-cell, development of the parameter estimation tool that was subsequently used by C. Peters.

Diffusion experiments are time-consuming and laborious (cf. Chapter 2). In the previous Chapter 5, optimal experimental designs for diffusion experiments are presented that allow for a reduction of the number of diffusion experiments without reducting the accuracy of diffusion coefficients. We showed that open geometries provide the most information on diffusion. However, with the current experimental setups, open geometries are not state of the art for diffusion measurements of fluids. The reason is that the boundary condition of a constant concentration at the opening of the geometry is difficult to realize in practice. In contrast, closed geometries are already practically implemented for diffusion measurements in fluids. For closed geometries, measurements at the wall provide most information on diffusion coefficients. Even better are spatially resolved measurements over time that enable a continuous tracking of the diffusion process (cf. Section 5.2).

In this chapter, model-based methods are applied in the development for two new closed geometry diffusion experiments with spatially resolved concentration measurements over time.
The first setup (Section 6.1) constists of a Loschmidt cell combined with interferometry. In the past, applications of this setup suffered from unknown systematic errors. To resolve the discrepancies, we perform a systematic model-based experimental analysis. The analysis provides the basis for improvements in the experimental setup and the model. The improved experiment is applied to measure concentration-dependent diffusion coefficients of binary gases from a only a single experimental run. Thereby, we achieve significant savings in experimental effort.
The second setup employs of a microfluidic device combined with Raman microspectroscopy. Here, a model-based experimental approach is applied in the development of a diffusion experiment that allows for the measurement of multicomponent liquid diffusion coefficients from a single experiment. The new experimental setup allows for a fast characterization of multicomponent diffusion in liquids with minimum sample consumption.

6.1 Concentration-dependent diffusion coefficients of binary gas mixtures from a single experiment

Gas diffusion experiments are complicated and time-consuming (cf. Chapter 2). Therefore, until today, great efforts are spent on the development of efficient gas diffusion experiments (Kugler et al., 2015b; Yue et al., 2018; Castañer and Ramírez, 2018; Medina and Ramírez, 2016; Kalyakin et al., 2018; Liu et al., 2016; Amalberti et al., 2018b). Due to high experimental effort, only a limited amount of reliable and accurate experimental data is available on full temperature-, pressure-, and composition-dependent behavior of gaseous diffusion coefficients.

In particular, only few experimental investigations analyzed the concentration dependence of diffusion coefficients in gases (Arora et al., 1978; Staker et al., 1975, 1974; van Heijningen et al., 1968). This is particularly challenging due to the weak dependency of diffusion coefficients on concentration in gases. One setup to measure binary diffusion coefficients in gases is the Loschmidt-cell combined with interferometry (Woolf et al., 1991; Winkelmann, 2007). The Loschmidt cell combined with interferometry has been applied successfully to obtain diffusion coefficients at fixed compositions (Marrero and Mason, 1972; Staker et al., 1974; Stewart et al., 1973; Gotoh et al., 1973; Boyd et al., 1951). However, multiple experiments are needed to obtain the concentration dependence of diffusion coefficients.

In principle, the concentration dependence can be obtained also from a single experiment by using pure gases. This idea was first proposed by Gupta and Cooper (Gupta and Cooper, 1971) and leads to the advantages that (1) fewer experiments are necessary, (2) less sample is consumed, and (3) that sample preparation is much easier since pure gases can be used instead of multiple pre-mixed gas mixtures which are difficult to prepare.

Up to now, the successful practical determination of concentration-dependent diffusion coefficients from a single experimental run has been limited to diffusion in liquids. Clunie et al. (Clunie et al., 1990) performed diffusion experiments in a diaphragm cell, but had large scatter in the results and still needed several experiments. Durou et al. (1974) performed diffusion experiments with Rayleigh (holographic) interferometry. Due to a limited precision of the measurements, good qualitative, but less good quantitative agreement with literature values was achieved. Bardow et al. (2005b) performed diffusion experiments with Raman spectroscopy and estimated reliable concentration-dependent diffusion coefficients in a mole fraction range from 20 to 85 mol-%. Kriesten et al. (2009) combined Raman spectroscopy and NMR measurements to obtain reliable concentration-dependent diffusion coefficients for the whole mole fraction range. Recently, Bouchaudy et al. (2018) introduced a microfluidic method for accurate measurements of concentration dependent diffusion coefficients in binary liquids.

Several attempts have been made to also obtain concentration-dependent diffusion coefficients of gases from single experiments performed in a Loschmidt cell combined with holographic interferometry (Kugler et al., 2013, 2015b; Kullnick, 2001; Baranski, 2002; Buttig et al., 2011; Kugler et al., 2015a). In all these attempts, the data showed systematic errors: The identified concentration dependencies were in disagreement with literature values and showed even diverging behavior for equimolar compositions and pure substances. Although a systematic study of Kugler et al. (2015a) could exclude many potential experimental error sources, the reason for the observed systematic errors could not be fully resolved.

To resolve the observed discrepancies, we perform a systematic experimental analysis. Thereby, an improved experimental setup, measurement procedure, and data evaluation procedure are developed that enable the determination of concentration-dependent binary gaseous diffusion coefficients from a single experiment. The new experimental setup and data analysis is validated with diffusion measurements of the system helium-krypton.

6.1.1 Experimental setup and procedure

The experimental setup consist of a Loschmidt cell combined with interferometry that was developed by Buttig et al. (Buttig et al., 2011; Buttig, 2010) and improved by Kugler et al. (Kugler et al., 2013, 2015a,b; Kugler, 2015). The Loschmidt cell corresponds to a closed diffusion cell as introduced in the previous Chapter 5 and in Figure 5.1(b). In the beginning of a diffusion experiment, the two half-cells of the Loschmidt cell are filled with binary gas mixtures of different composition. Removing a sliding plate between the two half-cells initiates the diffusion process. During the diffusion process, the index of refraction of the gas mixture inside the Loschmidt cell changes. The changes in the indices of refraction are measured by two interferometers, one for each the upper and lower half-cell. The interference patterns are recorded with cameras. Thus, compositions in the Loschmidt cell are measured spatially resolved over time.

The diffusion cell

Figure 6.1(b) illustrates the Loschmidt cell. The Loschmidt cell is made from steel and consists of two vertically aligned half-cells. The half-cells have windows that provide optical accessibility of the gas mixtures inside the Loschmidt cell. The half-cells can be connected or separated by a sliding plate in the middle of the Loschmidt cell. The sliding plate is moved magnetically such that no moving part has to be sealed against the ambient. Furthermore, glues and lubricants are avoided to avoid absorption in the glues or lubricants as well as contamination of the gases. The temperature inside the Loschmidt cell is controlled by a thermostatic housing operated with water. Thereby, stable temperatures within 0.01 K with a maximum offset of 0.025 K are obtained (Kugler, 2015). The Loschmidt cell is designed for pressures ranging from vacuum up to 1 MPa and for temperatures between 293.15 K and 353.15 K. The height of the Loschmidt cell is $L = 397.14\,\mathrm{mm}$, the depth (optical path) is $l = 201\,\mathrm{mm}$ (Kugler et al., 2013). More details on the construction of the Loschmidt cell are given by Kugler et al. (Kugler et al., 2013; Kugler, 2015).

The improved experimental setup

The optical setup consists of two identical interferometers, one interferometer for each half-cell of the Loschmidt cell. Figure 6.1(a) illustrates the optical setup of one of the half-cells. The beam of a He-Ne laser (1) is split into a reference beam and an object beam. The object beam passes the Loschmidt cell (5) and interferes with the reference beam on a holographic plate (7). The holographic plate is exposed to light and developed before the start of the diffusion experiment. Thereby, the hologram represents the original state inside the Loschmidt cell prior to the diffusion. The interference patterns are recorded every second with a camera (8). The positions of the interference fringes on the recorded images are converted to positions in the

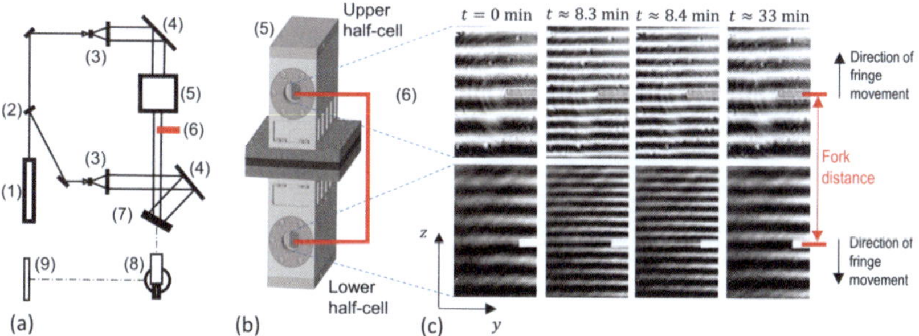

Figure 6.1: **(a)** Holographic interferometry setup used for each half-cell of the Loschmidt cell: The beam of a He-Ne laser (1) is split by a beam splitter (2) into the reference and the object beam. Both beams are expanded by the beam expander (3) and can be adjusted by two mirrors with piezo drives (4). The object beam passes the Loschmidt cell (5) and the calibration reference (6) and meets the reference beam on the surface of the hologram plate (7). Illumination of the hologram with the reference beam results in the reconstructed object beam which interferes with the (current) object beam that passes through the hologram plate. The interference of the reconstructed object beam and the current object beam is recorded with a camera (8). The camera is mounted on a rotational stage and is rotated towards the calibration scale (9) to obtain the calibration factors. The calibration reference (6) is used to correct possible camera orientation disturbances caused by the camera rotation.
(b) Loschmidt cell with the calibration scale in front of the windows. The calibration scale has the shape of a fork and is located in a way that each of its tines is in front of a window of the Loschmidt cell and enables a check and correction of the measurement calibration if necessary.
(c) Examples of recorded images of the upper and lower half-cell with interference patterns over time. The changing densities of the interference fringes over time show that the concentration gradient is increasing first ($t = 0\,\text{min}$ to $t = 8.4\,\text{min}$) and decreasing afterwards. The fringes move upwards in the upper half-cell and downwards in the lower half-cell as can be seen from the images at $t = 8.3\,\text{min}$ and $t = 8.4\,\text{min}$. The fork is visible in all images.

Loschmidt cell by the use of calibration factors.

The calibration factors are calculated from images taken from a calibration scale

(9). Before a diffusion experiment is performed, the camera is rotated towards the calibration scale to take an image. However, if the rotation mounts of the cameras are not aligned perfectly in parallel, the rotation of the cameras can lead to systematic offsets in the measurement locations. Although the spatial calibrations within the images of the upper and lower half-cell are correct, the distance between the images of the upper and lower half-cell would be systematically wrong. As a *first* improvement of the experimental setup, we introduced a calibration reference to check the camera's spatial calibration. The calibration reference has the shape of a fork. The distance of the fork's tines is known. Figure 6.1(b) illustrates how the fork (6) is installed in front of the Loschmidt cell (5) in such a way that the tines of the fork are in front of the windows of the upper and lower half-cell, respectively. Thereby, the tines of the fork are recorded together with the interference fringe patterns (cf. Figure 6.1(c)). Comparison of the camera's spatial calibration with the fork as a calibration reference revealed offsets in the measurement location of about 2 mm. In Section 6.1.3, we show the influence of these systematic errors and we develop an effective parameter estimation procedure to take care of these systematic errors.

When a diffusion experiment is performed, the half-cells of the Loschmidt cell are filled with binary gas mixtures of different composition. At the same time, the sliding plate is closed. The compositions of the gas mixtures in the upper and lower half-cell are homogeneous. Hence, there is no gradient in the index of refraction. Thus, the initial interference images, the so-called zero fields, have just only uniform intensities. In the past, the object beam was adjusted visually such that the zero field is uniformly bright with maximum intensity. The interference order of the zero field was assumed to be known. However, the visual setting of the zero field is difficult. Therefore, assuming a known interference order of the zero field can introduce significant errors in the data evaluation (cf. Section 6.1.3). Inspired by methods used to stabilize hologram fringes (MacQuigg, 1977; Neumann and Rose, 1967), our *second* improvement of the measurement procedure is the introduction of 6 to 15 initial interference fringes into the zero field by tilting the object or reference beam. Figure 6.1(c) shows examples of the recorded interference patterns over time for the upper and lower half-cell. In the beginning ($t = 0$ min), only the initial interference fringes are visible. To start the diffusion experiment, the sliding plate is opened. Thereby, the two half-cells of the Loschmidt cell are connected and the gas mixture compositions change depending on time and location in the Loschmidt cell. Hence, a concentration gradient arises. This concentration gradient is linked to a gradient in the index of refraction, which leads to the occurrence of new interference fringes. The resulting interference patterns are a superposition of the initial interference fringes and the interference fringes induced by diffusion. The interference fringes move upwards in the upper half-cell and downwards in the lower half-cell (cf. Figure 6.1(c) at $t = 8.3\,\text{min}$ and $t = 8.4\,\text{min}$). Towards the end of the diffusion process, the concentration equilibration concludes. Hence, the concentration gradient vanishes and thus the gradient in the index of refraction vanishes. Hence, the interference fringes induced by diffusion vanish (cf. Figure 6.1(c) at $t = 33\,\text{min}$).

The introduction of initial interference fringes into the zero field has several advantages:

1. The number of data points is increased significantly. In the data evaluation procedure, the positions and interference orders of the interference fringes are tracked as function of time (cf. Section6.1.2). When a bright zero field is used, the number of fringes is zero in the beginning, increases and decreases due to diffusion, and becomes zero again towards the end of the diffusion process. Hence, no interference fringes can be tracked in the beginning and after completion of the diffusion process. In contrast, when initial interference fringes are introduced, tracking of the interference fringes is possible from the beginning of the diffusion experiment and even after completion of the diffusion process. For a typical diffusion experiment with the system helium-krypton at 0.2 MPa with pure gases in the half-cells in the beginning of the diffusion experiment, we get approximately 3700 data points for the upper half-cell within the first 20 min (a data point is a fringe order number assigned to a measurement time and position). By using 6 initial interference fringes, the number of data points is doubled to approximately 8500 data points.

2. The movement of the interference fringes can be tracked more accurately. When a bright zero field is used, tracking of the interference fringes in the beginning and towards the end of the diffusion process is practically impossible, since single fringes stretch out over the total width of the recorded images. When initial interference fringes are used, the density of interference fringes on the recorded images is increased. The interference fringes are much more narrow, and narrow

interference fringes can be tracked more accurately. In particular, this applies to the beginning and the end of the diffusion process, when the number of fringes is small.

3. The stability of the optical setup can be evaluated. Since interference fringes can be tracked even after completion of the diffusion process, the stability of the optical setup can be evaluated by performing long-time experiments. In a perfectly stable setup, the interference fringes do not move before or after the diffusion process.

By use of initial interference fringes, we checked the stability of the optical setup. For this purpose, we performed an experiment where the Loschmidt cell is just filled with air. After setting the initial interference fringes, the positions of the interference fringes were tracked over time. From the movement of the interference fringes, the change in optical path length ΔOPL at one image position was calculated. At the same time, the temperature ϑ of the optical setup was recorded with a Pt-100 Ω resistance probe attached to the optical board of the lower half-cell. Changes in temperature of the optical setup lead to changes in the refractive index of air surrounding the optical setup and more important to thermally induced changes or movements of the optical boards and the optical equipment. Thereby, the optical path lengths of the object beams and the reference beams change and can lead to thermal drifts in the recorded interference fringe patterns. Figure 6.2(a) shows the changes in optical path length ΔOPL for the upper and lower half-cell and the temperature ϑ over time. Clearly, a correlation between ΔOPL and ϑ can be recognized. Changes in optical path length of up to 15 laser wavelengths could be detected for changes in temperature of 1 K. As a *third* improvement of the experimental setup, we installed a new thermal housing around the entire optical setup including the optical table to minimize thermal drifts. The thermal housing insulates the optical setup against the ambient temperature of the laboratory. The temperature inside the thermal housing is controlled by an air conditioner (Honeywell HCL-67E). Figure 6.2(b) illustrates the effect of the new temperature control on the optical stability. With temperature control, the temperature is held constant within 0.1 K leading to changes in optical path length of no more than 0.8 laser wavelengths. Although minimized, the very small, but remaining drifts have to be considered in the data evaluation procedure (cf. Section 6.1.2).

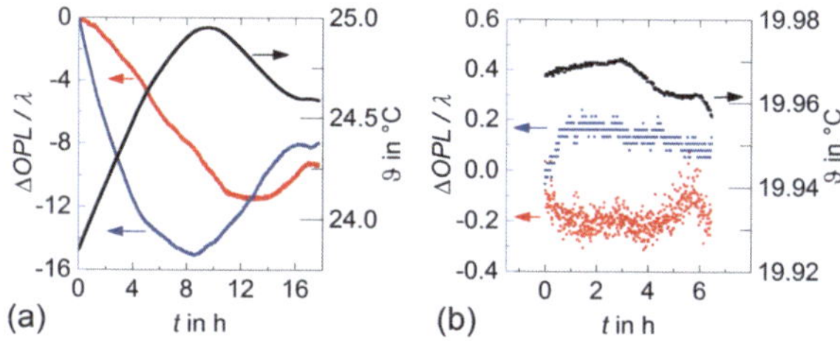

Figure 6.2: **(a)** Change of the optical pass length ΔOPL in wavelengths λ over time for the upper half-cell (red) and the lower half-cell (blue) when the temperature ϑ of the optical setup (black) outside of the Loschmidt cell is not controlled. The Loschmidt cell is filled with air only. **(b)** Change of the optical pass length ΔOPL in wavelengths λ over time for the upper half-cell (red) and the lower half-cell (blue) when the temperature ϑ of the optical setup (black) outside of the Loschmidt cell is controlled. Note the different scales on the y-axes.

6.1.2 Data evaluation

The experimental data consists of a time series of images with interference patterns as shown exemplary in Figure 6.3(c). In this section, we explain how the images are evaluated to obtain experimental interference fringe orders as a function of time and position. Subsequently, we present the diffusion model, which relates the diffusion coefficient to interference fringe orders. Finally, the parameter estimation procedure is introduced to relate the experimental interference fringe orders to the diffusion model to obtain diffusion coefficients.

Image processing

Processing of the images consists of 4 steps: In the 1st step, the intensities of the 2-dimensional images are averaged in the direction perpendicular to diffusion, the y-direction. Thereby, we obtain 1-dimensional intensity profiles with interference patterns. Figure 6.3(a) shows the intensity profile of the initial interference pattern shown in Figure 6.1(c) at $t = 0\,\text{min}$ for the upper half-cell.

In the 2nd step, the positions $\tilde{z}_{\text{meas}}$ of the interference fringe minima of each 1-dimensional intensity profile are identified (cf. the blue dashed lines in Figure 6.3(a)).

Considering the movement direction of the interference fringes (upwards in the upper half-cell and downwards in the lower half-cell), the interference fringe minima are numbered consecutively starting from zero. Since the measurement time t_{meas} of each recorded image is also known, we thereby obtain the interference fringe orders $\tilde{k}_{\text{meas}}(\tilde{z}_{\text{meas}}, t_{\text{meas}})$ as a function of measurement positions $\tilde{z}_{\text{meas}}$ and measurement times t_{meas} (cf. the blue stars in Figure 6.3(b)).

In the 3rd step, the influence of the initial interference fringes on the numbering of the interference fringe orders is removed. For this purpose, the initial interference fringe orders $\tilde{k}_{\text{meas}}(\tilde{z}_{\text{meas}}, t_{\text{meas}} = 0)$ (cf. the red line in Figure 6.3(b)) are subtracted from all interference fringe orders $\tilde{k}_{\text{meas}}(z_{\text{meas}}, t_{\text{meas}})$:

$$k_{\text{meas}}(\tilde{z}_{\text{meas}}, t_{\text{meas}}) = \tilde{k}_{\text{meas}}(\tilde{z}_{\text{meas}}, t_{\text{meas}}) - \tilde{k}_{\text{meas}}(\tilde{z}_{\text{meas}}, t_{\text{meas}} = 0) \tag{6.1}$$

(cf. the black stars in Figure 6.3(b)).

In the 4th step, we account for the camera calibration offsets (cf. Section 6.1.1). From the recorded images (cf. Figure 6.1(c)), we calculate the distance of the fork's tines, $\Delta z_{\text{fork,camera}}$, based on the camera calibration factors. The difference between the known distance of the fork's tines, Δz_{fork}, and $\Delta z_{\text{fork,camera}}$ is the camera offset

$$\Delta z_{\text{camera}} = \Delta z_{\text{fork}} - \Delta z_{\text{fork,camera}}. \tag{6.2}$$

The camera offset could be either added to the measured interference fringe positions of the upper half-cell or subtracted from the measured interference fringe positions of the lower half-cell. We chose to add the camera offset to the measured interference fringe positions of the upper half-cell. Thereby, we obtain the measurement positions

$$z_{\text{meas}} = \begin{cases} \tilde{z}_{\text{meas}} + \Delta z_{\text{camera}} & \text{(upper half - cell)} \\ \tilde{z}_{\text{meas}} & \text{(lower half - cell)} \end{cases} \tag{6.3}$$

Thereby, the relative distances between the measurement positions z_{meas} are corrected. However, the absolute values of the measurement positions are still associated with uncertainty. We account for the uncertainties in the absolute values of the measurement positions in the parameter estimation procedure (see below).

Finally, the resulting interference fringe orders $k_{\text{meas}}(z_{\text{meas}}, t_{\text{meas}})$ form the raw data for the parameter estimation procedure. An example for the interference fringe orders $k_{\mathrm{meas}}(z_{\mathrm{meas}}, t_{\mathrm{meas}})$ is shown in Figure 6.3(c).

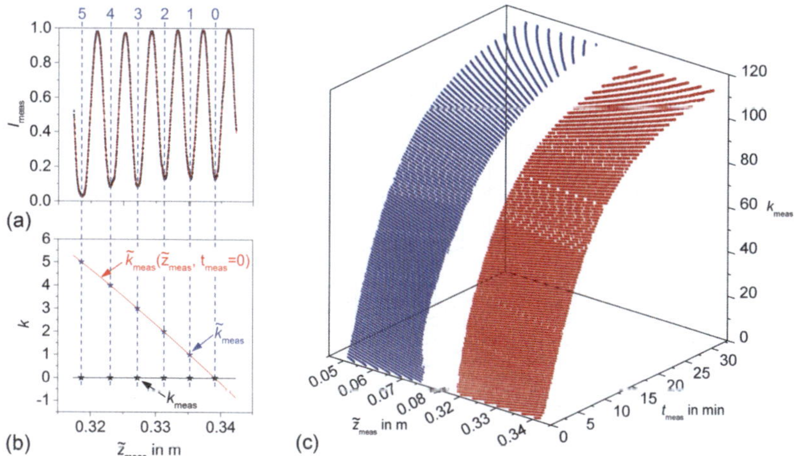

Figure 6.3: **(a)** Intensity profile in z-direction of the initial interference pattern shown in Figure 6.1(c) at time $t = 0\,\mathrm{min}$ for the upper half-cell. The intensities are averaged in the direction perpendicular to diffusion, the y-direction (cf. Figure 6.1(c)). Black: Measured intensities. Red: Smoothed intensity profile. Blue: Identified minima of the smoothed intensity profile with corresponding interference fringe orders $\tilde{k}_{\mathrm{meas}}$. Since the interference fringes move upwards in the upper half-cell, the interference fringe orders $\tilde{k}_{\mathrm{meas}}$ are numbered consecutively opposite to the z-direction. **(b)** Interference fringe orders as a function of position $\tilde{z}_{\mathrm{meas}}$ of the same initial interference pattern. Blue: Consecutively numbered interference orders $\tilde{k}_{\mathrm{meas}}$. Black: Corrected interference fringe orders k_{meas} after subtraction of the initial interference fringe orders $\tilde{k}_{\mathrm{meas}}(\tilde{z}_{\mathrm{meas}}, t_{\mathrm{meas}} = 0)$. For the special case of the initial interference pattern, the corrected initial interference fringe orders are $k_{\mathrm{meas}}(\tilde{z}_{\mathrm{meas}}, t_{\mathrm{meas}} = 0) = 0$ by definition. After start of the diffusion process, the interference fringe orders $k_{\mathrm{meas}}(\tilde{z}_{\mathrm{meas}}, t_{\mathrm{meas}})$ increase with time (cf. adjacent Figure 6.3(c)). **(c)** Interference fringe orders k_{meas} as function of location $\tilde{z}_{\mathrm{meas}}$ and time t_{meas}. Red: upper half-cell, blue: lower half-cell.

Diffusion process model

For large initial composition differences, we cannot assume ideal gas behavior, but have to account for the real gas behavior and include excess effects. The partial molar volumes change during the diffusion process; as a result, the total volume of the diffusing gases would change, but the Loschmidt cell has a constant volume. Therefore, the pressure in the Loschmidt cell changes during the diffusion process. The center of mass moves which requires to include a convective velocity in the differential mass balance. However, the resulting differential mass balance would be difficult to solve using standard simulation software.

In contrast to the pressure in the Loschmidt cell, the total number of moles is constant during the diffusion process. Therefore, it is more convenient to use the molar reference frame (cf. Section 2.1.2). The 1-dimensional mass balance combined with Fick's first law then reads as (cf. Equation 2.23)

$$\frac{\partial \xi_1(\bar{z},t)}{\partial t} = \frac{\partial}{\partial \bar{z}} \left(D\,(c_t(\bar{z},t)\,V_2^0)^2 \,\frac{\partial \xi_1(\bar{z},t)}{\partial \bar{z}} \right). \tag{6.4}$$

We describe the concentration dependence of the diffusion coefficient by the linear relation

$$D = \left[D^{(1)} \cdot x_1 + D^{(2)} \cdot (1 - x_1)\right] \cdot \frac{0.2\,\mathrm{MPa}}{p}. \tag{6.5}$$

The term $0.2\,\mathrm{MPa}/p$ accounts for the pressure dependence of the diffusion coefficient: $D \propto 1/p$ (cf. Section 2.1.3). Thereby, $D^{(1)}$ and $D^{(2)}$ are diffusion coefficient parameters at a pressure of $0.2\,\mathrm{MPa}$. Other models for the diffusion coefficient could be inserted in the model if required. The molar concentrations and mole fractions are related by

$$c_t(z,t) = c_1(z,t) + c_2(z,t), \quad x_1 = \frac{c_1(z,t)}{c_t(z,t)}. \tag{6.6}$$

c_1 and c_2 are related through the pressure virial equation (Gmehling et al., 2012)

$$\begin{aligned}\frac{p(t)}{RT} =& c_1(z,t) + c_2(z,t)\\ &+ B_{11}\,c_1^2(z,t) + 2B_{12}\,c_1(z,t)c_2(z,t) + B_{22}\,c_2^2(z,t)\\ &+ C_{111}\,c_1^3(z,t) + 3C_{112}\,c_1^2(z,t)\,c_2(z,t) + 3C_{122}\,c_1(z,t)\,c_2^2(z,t) + C_{222}\,c_2^3(z,t).\end{aligned} \tag{6.7}$$

A further constraint for the pressure p and the molar concentrations c_t, c_1, and c_2 is given by the constant length L of the Loschmidt cell

$$L = \int_0^{\bar{L}} \frac{1}{c_t(\bar{z}, t)\, V_2^0}\, d\bar{z}' = \text{const.} \neq f(t) \tag{6.8}$$

which follows from Equation 2.20.

The diffusion mole balances have to be complemented by additional initial conditions, boundary conditions, and constitutive equations. The concentrations c_i are related to the index of refraction of the mixture, n, by the Lorentz-Lorenz equation

$$\frac{n^2(z,t) - 1}{n^2(z,t) + 2} = A_{R,1}\, c_1(z,t) + A_{R,2}\, c_2(z,t). \tag{6.9}$$

Here, $A_{R,1}$ and $A_{R,2}$ are the first refractivity virial coefficients of the pure components 1 and 2. For the system helium-krypton, the first refractivity coefficients are $A_{R,He} = 0.5213\,\text{cm}^3/\text{mol}$ and $A_{R,Kr} = 6.414\,\text{cm}^3/\text{mol}$ (Achtermann et al., 1993).

The difference $\Delta n(z,t)$ between the indices of refraction at time t and at the beginning of the experiment,

$$\Delta n(z,t) = n(z,t) - n(z,t=0), \tag{6.10}$$

leads to differences in the optical path lengths between the object beam and the reconstructed object beam

$$\Delta L_{opt}(z,t) = l\,\Delta n(z,t). \tag{6.11}$$

Here, $l = 201\,\text{mm}$ is the depth of the Loschmidt cell (Kugler et al., 2013), which corresponds to the geometrical path length of the laser beam through the Loschmidt cell. Diffusion-induced interference fringes with order

$$k_{diff}(z,t) = \frac{\Delta L_{opt}(z,t)}{\lambda} \tag{6.12}$$

occur, if $\Delta L_{opt}(z,t)$ is an integer multiple of the laser wavelength λ. A He-Ne laser operated at a wavelength of $\lambda = 632.8\,\text{nm}$ was used (Kugler et al., 2013).

In addition to diffusion, thermal instabilities of the optical setup can change the interference fringe orders (cf. Section 6.1.1). To account for the remaining small but still noticeable linear drifts in temperature, we allow for linear drifts $k_{\text{drift}} \cdot t$ in the interference fringe orders. Furthermore, the initial interference fringe orders k_0 that remain after subtraction of the initial interference fringes (cf. Equation 6.1) have to be considered in the calculation of the observable interference fringe order $k(z,t)$. The observable interference fringe order is therefore

$$k(z,t) = \begin{cases} k_{0,\text{upper}} + k_{\text{drift,upper}} \cdot t + k_{\text{diff}}(z,t) & \text{for } z \geq z_0 \text{ (upper half - cell)} \\ k_{0,\text{lower}} + k_{\text{drift,lower}} \cdot t + k_{\text{diff}}(z,t) & \text{for } z < z_0 \text{ (lower half - cell)} \end{cases} \tag{6.13}$$

Here, $k_{0,\text{upper}}$ and $k_{0,\text{lower}}$ are the initial values of the interference fringe orders in the upper and lower half-cell. z_0 is the coordinate of the initial domain boundary between upper and lower half-cell. The terms $k_{\text{drift,upper}} \cdot t$ and $k_{\text{drift,lower}} \cdot t$ account for linear drifts of the interference fringe orders in the upper and lower half-cell.

The model is complemented by initial and boundary conditions. The initial composition profile is

$$x_1(z,t=0) = \begin{cases} x_{1,0,\text{upper}} & \text{for } z \geq z_0 \text{ (upper half - cell)} \\ x_{1,0,\text{lower}} & \text{for } z < z_0 \text{ (lower half - cell)} \end{cases} \tag{6.14}$$

Here, $x_{1,0,\text{upper}}$ and $x_{1,0,\text{lower}}$ are the initial gas mixture compositions in the upper and lower half-cell, respectively. The initial pressure is given by

$$p(t=0) = p_0. \tag{6.15}$$

The initial indices of refraction $n(z,t=0)$ are calculated with the Lorentz-Lorenz equation (Equation 6.9) by use of the known initial compositions $x_1(z,t=0)$.

The boundary conditions are given by the no-flux conditions

$$\left(\frac{\partial x_1(z,t)}{\partial z}\right)_{z=0} = \left(\frac{\partial x_1(z,t)}{\partial z}\right)_{z=L} = 0 \tag{6.16}$$

at the lower ($z=0$) and upper ($z=L$) walls of the Loschmidt cell.

Parameter estimation

The interference fringe orders k_{meas} obtained from the recorded interference patterns are used as measurement data. We assume a structural exact model and a normally distributed measurement error with constant standard deviation $\sigma_{k_{\text{meas}}}$. Therefore, unweighted least squares are used for the parameter estimation (cf. Section 3.3). The parameters are obtained from the minimization problem

$$\Phi_{\min} = \min_{\boldsymbol{\theta}} \sum_{\mu=1}^{N_{\text{exp}}} \left(\boldsymbol{k}_{\mu}(\boldsymbol{z}_{\text{meas}} + \boldsymbol{\Delta z}_{\text{corr}}, \boldsymbol{t}_{\text{meas}} + \boldsymbol{t}_0) - \boldsymbol{k}_{\text{meas},\mu} \right)^2 , \tag{6.17}$$

where N_{exp} denotes the number of experimental repetitions. $\boldsymbol{k}_{\mu}(\boldsymbol{z}_{\text{meas}} + \boldsymbol{\Delta z}_{\text{corr}}, \boldsymbol{t}_{\text{meas}} + \boldsymbol{t}_0)$ are the vectors of interference fringe orders calculated from the diffusion model at positions $\boldsymbol{z}_{\text{meas}} + \boldsymbol{\Delta z}_{\text{corr}}$ and at times $\boldsymbol{t}_{\text{meas}} + \boldsymbol{t}_0$. The parameters $\boldsymbol{\theta}$ to be estimated are

$$\boldsymbol{\theta} = \left[D^{(1)}, D^{(2)}, \boldsymbol{z}_0, \boldsymbol{t}_0, \boldsymbol{p}_0, \boldsymbol{k}_{0,\text{upper}}, \boldsymbol{k}_{0,\text{lower}}, \boldsymbol{k}_{\text{drift,upper}}, \boldsymbol{k}_{\text{drift,lower}}, \boldsymbol{\Delta z}_{\text{corr}} \right] \tag{6.18}$$

and explained in more detail in the following. All estimated parameters besides the diffusion coefficients $D^{(1)}$ and $D^{(2)}$ are vectors because they are estimated for each experiment individually. All estimated parameters besides the diffusion coefficients $D^{(1)}$ and $D^{(2)}$ are nuisance parameters which are not of immediate interest but have to be accounted for to get statistically sound results:

- The position z_0 of the initial domain boundary between the upper and lower half-cell is fitted to take care of possible distortions introduced by opening the sliding plate in the beginning of the diffusion experiment.
- The time offset t_0 between the start of the measurement and the beginning of the diffusion process is fitted because the actual start of the diffusion process is known only with limited precision due to the time needed to open the sliding plate in the beginning of the diffusion experiment (Fujita, 1956).
- The initial pressure p_0 is fitted to compensate for calibration uncertainty of the pressure sensor.
- The initial fringe orders $k_{0,\text{upper}}$ and $k_{0,\text{lower}}$ are fitted because they are known only with limited precision. The opening of the sliding plate in the beginning of the diffusion experiment leads to small distortions and a slight offset of the initial fringes. In Section 6.1.3, we demonstrate the high impact of the initial

fringe order k_0 on the results of the experimental evaluation. The necessity to fit the initial fringe order has also been found by Gotoh et al. (1973). However, fitting the initial fringe order k_0 has not been applied in the previous works of Kullnick (2001), Baranski (Baranski, 2002), Buttig (Buttig et al., 2011), and Kugler et al. (2015a).

- The linear drift parameters $k_{\text{drift,upper}}$ and $k_{\text{drift,lower}}$ are fitted to account for the remaining small but still noticeable linear drifts in temperature during the experiment (cf. Section 6.1.1). We demonstrate the effect of linear drifts on the results of the experimental evaluation in Section 6.1.3.
- The measurement location correction $\Delta \boldsymbol{z}_{\text{corr}}$ is fitted to account for spatial offsets in the measurement locations z_{meas} due to camera calibration offsets (cf. Section 6.1.1). The rotation of the cameras from the calibration scale towards the diffusion cell can lead to offsets of about 2 mm in the relative distances between the measurement position of the upper and lower half-cell. We corrected this offset by using the calibration reference (cf. Section 6.1.1). Thereby, the relative distances between the measurement positions z_{meas} are corrected, but the absolute values of the measurement positions are still associated with uncertainty. Therefore, we introduced the fitting parameter $\boldsymbol{\Delta} z_{\text{corr}}$, which allows for shifting of the measurement positions while the relative distances of the measurement positions between upper and lower half-cell are kept constant. We demonstrate the effect of spatial offsets in Section 6.1.3.

6.1.3 Identification of necessary improvements in experiment and model by model-based experimental analysis

The aim of the work presented in this chapter is to determine concentration-dependent binary diffusion coefficients of nobel gases from a single experimental run. For this purpose, the long-standing discrepancies in connection with the application of the Loschmidt cell technique have to be resolved.

We applied the model-based experimental analysis introduced in Section 3.3 to identify the necessary improvements of the experimental setup (cf. Section 6.1.1) and the diffusion model (cf. Section 6.1.2). For this purpose, we analyzed the statistical properties of the residuals $k - k_{\text{meas}}$ of the interference fringe orders after the parameter estimation. The residuals should be normally distributed and should show no time-dependent behavior. Furthermore, the residuals have to be below a certain threshold value to identify the concentration dependence of the diffusion coefficient. To evaluate this threshold value, we calculated interference patterns with our diffusion model for concentration-dependent diffusion coefficients of the system helium-krypton taken from literature (cf. Figure 6.6). To simulate a real experiment, we added a typical image noise from our measurements (signal-to-noise ratio = 30). Subsequently, we estimated a constant diffusion coefficient from the calculated interference patterns. Thereby, maximum residuals $k - k_{\text{meas}} = 0.2$ were observed. Thus, systematic measurement errors have to be below this threshold of 0.2; residuals $k - k_{\text{meas}} < 0.2$ are required to identify concentration-dependent diffusion coefficients of the system helium-krypton. In diffusion experiments with pure gases, 120 interference fringe orders are observed in total (cf. Figure 6.3(c)). Thus, the acceptable residuals $k - k_{\text{meas}} < 0.2$ correspond to relative errors of $0.2/120 = 0.17\,\%$ between experimental data and diffusion model, emphasizing the need for highly accurate diffusion measurements.

In the following, we demonstrate the impact of offsets in the initial fringe order k_0 on the experimental evaluation. In addition, we demonstrate the effect of small linear drifts caused by thermal drifts of the optical setup (cf. Section 6.1.1) on the results of the experimental evaluation. Subsequently, we demonstrate the effect of spatial offsets in the measurement locations, which motivated the installation of a calibration reference (cf. Section 6.1.1) and the inclusion of the fitting parameter Δz_{corr} (cf. Section 6.1.2).

Impact of offsets in the initial fringe order k_0

The visual setting of a bright zero field in the beginning of a diffusion experiment and the opening of the sliding plate in the beginning of a diffusion experiment leads to small distortions and a slight offset of the initial fringes. Therefore, the initial interference fringe orders $k_{0,\text{upper}}$ and $k_{0,\text{lower}}$ of the upper and the lower half-cell contain uncertainty and have to be included in the parameter estimation (cf. Section 6.1.2).

To demonstrate the exclusive impact of the initial fringe order on the residuals without impact of other nuisance parameters and without impact of concentration dependencies and real gas effects, we performed a simulation study for diffusion experiments with small initial composition differences of 30 mol-% between the gas mixtures in the upper and lower half-cell of the Loschmidt cell. The concentration dependence of the diffusion coefficient is very weak: For the system helium-krypton, the diffusion coefficients differ by at most 5 % for arbitrary compositions and by at most 2 % within 30 mol-% intervals. Therefore, we can assume an ideal mixture of ideal gases with virial coefficients $B_{ij} = C_{ijk} = 0$ and with a constant diffusion coefficient $D^{(1)} = D^{(2)} = D = 3.14 \cdot 10^{-5}\,\mathrm{m^2/s}$ within the initial composition differences of 30 mol%. Preliminary calculations showed that the maximum error in the diffusion coefficient introduced by the assumption of ideal gas behavior and a constant diffusion coefficient is 0.04 %, which is orders of magnitudes below the concentration dependence.

Figure 6.4(a) shows the residuals of a simulation of a perfect diffusion experiment. The image noise of the recorded interference patterns with a signal-to-noise ratio of 30 leads to a root-mean-square error of the residuals of $RMSE = 0.024$. The initial fringe orders in the upper and lower half-cell are $k_{0,\mathrm{upper}} = k_{0,\mathrm{lower}} = 0.5$.

Figure 6.4(b) shows the residuals when typical initial fringe orders of $k_{0,\mathrm{upper}} = 0.83$ and $k_{0,\mathrm{lower}} = 0.67$ are applied but are not considered in the parameter estimation. The residuals show significant time-dependence, i.e., a time-dependence that is of the order of magnitude that is important to retrieve the concentration dependence of the diffusion coefficient. The $RMSE$ of the residuals is 0.118. The estimated diffusion coefficient is $D = 3.28 \cdot 10^{-5}\,\mathrm{m^2/s}$. Hence, the estimated diffusion coefficient deviates by 4.3 % from the assumed true value.

Impact of time-dependent linear drifts of the interference fringes k_{drift}

The temperature control of the optical setup guarantees stable temperatures within 0.1 K (cf. Section 6.1.1). However, even slight linear drifts in temperature in the range of 0.1 K can lead to temporal drifts of the interference fringe orders that are of the order of magnitude that is important to retrieve the concentration dependence of the diffusion coefficient. Figure 6.4(c) shows the residuals of the simulated diffusion experiment introduced in Section 6.1.3/Figure 6.4(a), when typical linear drifts $k_{\mathrm{drift,upper}} = 0.022/\mathrm{min}$ for the upper half-cell and $k_{\mathrm{drift,lower}} = -0.026/\mathrm{min}$ for the lower half-cell are applied. These drifts relate to a temperature drift of approximately $T_{\mathrm{drift}} = 3 \cdot 10^{-3}\,\mathrm{K/min}$. The residuals show significant time-dependence. The $RMSE$ of the residuals is 0.080. Due to the counteracting drifts in the upper and lower half-cell

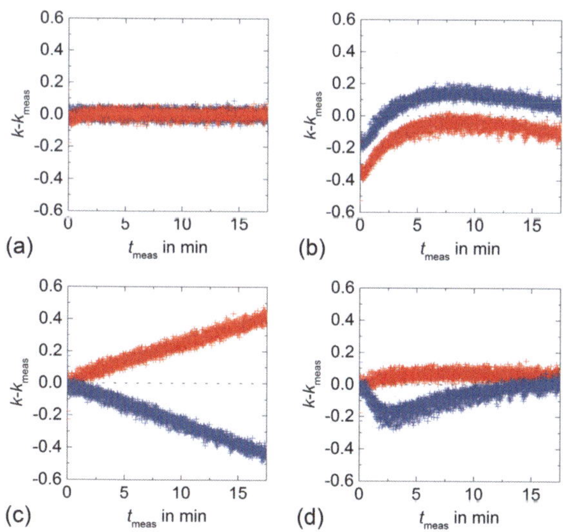

Figure 6.4: Residuals $k - k_{\mathrm{meas}}$ over time t_{meas} of a simulated diffusion experiment of the system helium-krypton at 293.15 K and 0.2 MPa when only the diffusion coefficient D is fitted. The initial compositions are $x_{\mathrm{Kr,0,upper}} = 0.3$ in the upper half-cell and $x_{\mathrm{Kr,0,lower}} = 0.6$ in the lower half-cell. A typical image noise of the recorded interference patterns with a signal-to-noise ratio of 30 was applied. Red: residuals for the upper half-cell; blue: residuals for the lower half-cell. **(a)** Residuals of a perfect diffusion experiment with initial fringe orders $k_{0,\mathrm{upper}} = k_{0,\mathrm{lower}} = 0.5$ for the upper and lower half-cell. **(b)** Residuals of a diffusion experiment affected by offsets in the initial fringe orders: $k_{0,\mathrm{upper}} = 0.83$ for the upper half-cell and $k_{0,\mathrm{lower}} = 0.67$ for the lower half-cell. **(c)** Residuals of a diffusion experiment affected by linear drifts in the interference fringe orders induced by thermal drifts of the optical setup, $k_{\mathrm{drift,upper}} = 0.022/\mathrm{min}$ for the upper half-cell and $k_{\mathrm{drift,lower}} = -0.026/\mathrm{min}$ for the lower half-cell. These drifts relate to a temperature drift of approximately $T_{\mathrm{drift}} = 3 \cdot 10^{-3}\,\mathrm{K/min}$. **(d)** Residuals of a diffusion experiment affected by a spatial offset of $\Delta z = 2\,\mathrm{mm}$ in the relative distances between the measurement positions of the upper and lower half-cell.

leading to almost symmetric residuals of the upper and lower half-cell, the estimated diffusion coefficient $D = 3.13 \cdot 10^{-5}\,\mathrm{m^2/s}$ matches the assumed true value accidentally.

Impact of spatial offsets

The rotation of the cameras from the calibration scale towards the diffusion cell can lead to spatial offsets of about 2 mm in the relative distances between the measurement positions of the upper and lower half-cell (cf. Section 6.1.1). Figure 6.4(d) shows the residuals of the simulated diffusion experiment introduced in Section 6.1.3/Figure 6.4(a), when a typical spatial offset $\Delta z = 2\,\mathrm{mm}$ in the measurement positions of the upper half-cell is applied. The residuals show significant time-dependence. The *RMSE* of the residuals is 0.146. The estimated diffusion coefficient is $D = 3.16 \cdot 10^{-5}\,\mathrm{m^2/s}$, and, thus, deviates by 1.3 % from the assumed true value.

6.1.4 Validation of the experimental setup: Concentration-dependent diffusion coefficients from multiple experiments

The improved measurement setup and evaluation procedure are now applied to determine concentration-dependent diffusion coefficients of the system helium-krypton at 293.15 K and 0.2 MPa. In a first step, the concentration dependence is determined from multiple experiments with small composition differences between the lower and upper half-cell. Thereby, the concentration-dependence of the diffusion coefficient and real gas effects are excluded and full attention can be paid to the validation of the experimental setup.

The choice of the composition difference is based on a trade-off. On the one hand, a small composition difference leads to a better approximation of the concentration dependence of the diffusion coefficient $D(x)$. On the other hand, a small composition difference leads also to a less accurate estimation of the averaged diffusion coefficient, since less interference fringes occur during the diffusion process. We chose a composition difference of 30 mol% where up to 34 fringes occur during the diffusion process. To obtain the concentration dependence, we chose 4 different mixtures with total compositions ranging from krypton mole fractions of $x_{\mathrm{Kr}} = 0.15$ to $x_{\mathrm{Kr}} = 0.85$. The gas mixtures provided by Linde® are pre-mixed with relative molar uncertainties of 1 %. Each experiment is repeated at least 4 times (cf. Table 6.1).

The diffusion coefficients are obtained from the least-squares optimization according

Table 6.1: Compositions and number of repetitions of the diffusion experiments

Mean mole fraction of Kr	Mole fraction of Kr in the lower half cell	Mole fraction of Kr in the upper half-cell	Number of repetitions
0.15	0.303	0.00001	6
0.45	0.608	0.303	7
0.55	0.707	0.404	4
0.85	0.99999	0.706	6

to Equation 6.17. Herein, all repeated experiments are evaluated simultaneously to improve the statistical quality of the results. In addition to the diffusion coefficient, the full set of nuisance parameters is fitted, i.e., the initial conditions z_0, t_0, p_0, $k_{0,\text{upper}}$, the drift corrections $k_{\text{drift,upper}}$, $k_{\text{drift,lower}}$, and the correction of the measurement location offset z_{corr} (cf. Section 6.1.2).

Figure 6.5 shows the residuals of an experiment with an overall mole fraction of krypton of $x_{\text{Kr}} = 0.45$. Figure 6.5(a) shows the residuals when only the diffusion coefficient D is fitted. The distinct time dependence indicates that the fitted diffusion coefficient $D = 2.99 \cdot 10^{-5}\,\text{m}^2/\text{s}$ is inaccurate, which is indicated by an $RMSE$ of 0.169. Furthermore, the residuals exceed the threshold of 0.2 that is required to identify concentration-dependent diffusion coefficients of the system helium-krypton.

Figure 6.5(b) shows the residuals, when the diffusion coefficient D and the nuisance parameters t_0, z_0, and p_0 are fitted ($D = 3.14 \cdot 10^{-5}\,\text{m}^2/\text{s}$, $t_0 = 8.1\,\text{s}$, $z_0 - L/2 = 1.9\,\text{mm}$, and $p_0 = 0.197\,\text{MPa}$). The time dependence of the residuals is diminished, but still present ($RMSE = 0.07$).

Figure 6.5(c) shows the residuals, when the diffusion coefficient D and the full set of nuisance parameters t_0, z_0, p_0, $k_{0,\text{upper}}$, $k_{0,\text{lower}}$, $k_{\text{drift,upper}}$, $k_{\text{drift,lower}}$, and Δz_{corr} are fitted ($D = 3.14 \cdot 10^{-5}\,\text{m}^2/\text{s}$, $t_0 = 0.3\,\text{s}$, $z_0 - L/2 = -0.7\,\text{mm}$, $p_0 = 2.02\,\text{MPa}$, $k_{0,\text{upper}} = 0.42$, $k_{0,\text{lower}} = 0.52$, $k_{\text{drift,upper}} = -5.81 \cdot 10^{-4}/\text{min}$, $k_{\text{drift,lower}} = 4.83 \cdot 10^{-2}/\text{min}$, and $\Delta z_{\text{corr}} = -2.2\,\text{mm}$. With the full set of nuisance parameters, the residuals show no time-dependence and the residuals are well below the threshold of 0.2 that is required to identify concentration-dependent diffusion coefficients of the system helium-krypton ($RMSE = 0.04$).

Figure 6.6 shows the diffusion coefficients over the mean mole fraction of krypton x_{Kr} for our experiments (red) in comparison to literature values (Carson and Dun-

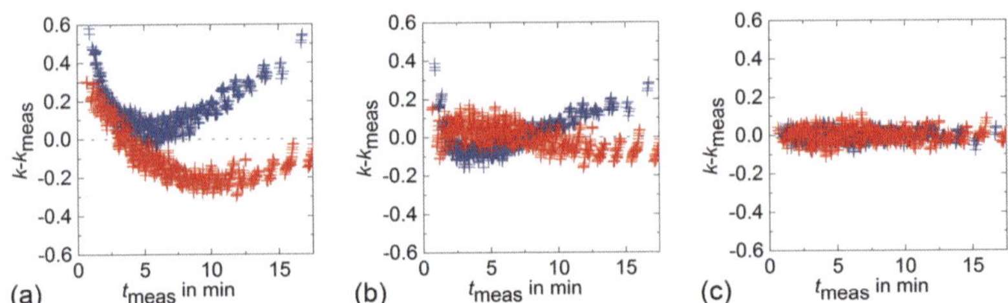

Figure 6.5: Residuals $k - k_{\mathrm{meas}}$ over time t_{meas} of a real diffusion experiment of the system helium-krypton at 293.15 K and 0.2 MPa. The initial compositions are $x_{\mathrm{Kr,0,upper}} = 0.3$ in the upper half-cell and $x_{\mathrm{Kr,0,lower}} = 0.6$ in the lower half-cell. Red: residuals for the upper half-cell; blue: residuals for the lower half-cell. **(a)** Residuals when only the diffusion coefficient D is fitted. **(b)** Residuals when the diffusion coefficient D and the nuisance parameters t_0, z_0, and p_0 are fitted. **(c)** Residuals when the diffusion coefficient D and the full set of nuisance parameters t_0, z_0, p_0, $k_{\mathrm{0,upper}}$, $k_{\mathrm{0,lower}}$ $k_{\mathrm{drift,upper}}$, $k_{\mathrm{drift,lower}}$, and z_{corr} are fitted. The full set of nuisance parameters is necessary to obtain residuals without time-dependence.

lop, 1972; Srivastava and Paul, 1962; Kestin et al., 1984; Hogervorst, 1971; Staker and Dunlop, 1976; Staker et al., 1974) (black) (numerical values can be found in Appendix C). The error bars of our own measurements represent 95 % confidence intervals (cf. Equation 3.35). If necessary, the literature values have been corrected to 293.15 K and 0.2 MPa with the relations $D \propto T^{11/6}$ (Marrero and Mason, 1972) and $D \propto 1/p$ (Cussler, 2007) (the original temperature and pressure conditions of the literature experiments are given in Appendix C). Overall, our fitted diffusion coefficients agree well with the literature values. Only the low mole fraction range $0 \leq x_{\mathrm{Kr}} \leq 0.25$ seems to be ambiguous. Here, the deviations are outside combined uncertainties of our data and those from Carson and Dunlop (1972). However, our data is well in line with the data from Kestin et al. (1984), Hogervorst (1971), Staker and Dunlop (1976), and Srivastava and Paul (1962).

Overall, the improved measurement setup and evaluation procedure enables the successful determination of the concentration-dependent diffusion coefficient of the system helium krypton from multiple experiments.

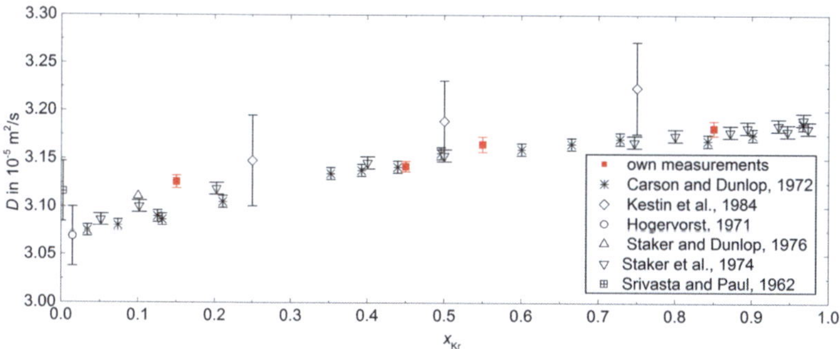

Figure 6.6: Experimental diffusion coefficients for the system helium-krypton as function of the mole fraction of krypton x_{Kr} at 293.15 K and 0.2 MPa. Our measurement data (red) represents diffusion coefficients averaged over a mole fraction range of 30 mol%. The error bars of our measurements (red) represent 95 % confidence intervals. Numerical values of the diffusion coefficients are given in Appendix C. The literature data has been corrected to 293.15 K and 0.2 MPa with the relations $D \propto T^{11/6}$ (Marrero and Mason, 1972) and $D \propto 1/p$ (Cussler, 2007) if necessary.

6.1.5 Concentration-dependent diffusion coefficients from single experimental runs

After validation, the improved measurement setup and evaluation procedure was applied to determine concentration-dependent diffusion coefficients of the system helium-krypton from a single experiment at 293.15 K and 0.2 MPa. In the beginning of a measurement, the lower half-cell was filled with pure krypton and the upper half-cell was filled with pure helium.

For experiments with pure gases, taking care of thermal drifts in the optical setup is even more important than for measurements with small composition differences. Therefore, the measurements with pure gases were performed for 6 hours: The data of the first 33 minutes are used for the diffusion coefficient evaluation; the data after 33 minutes are used to evaluate the stability of the optical setup. Only if the interference fringes do not move or move only approximately linearly with time, the measurement data is used for the evaluation of diffusion coefficients. In total, 6 measurements with acceptable thermal stability after completion of the diffusion process were performed.

In order to identify concentration-dependent diffusion coefficients of the system helium-krypton, residuals $k - k_{\text{meas}} < 0.2$ are required (cf. Section 6.1.3). To identify experimental runs satisfying the quality criterion of residuals $k - k_{\text{meas}} < 0.2$, we first performed an individual evaluation of each run, i.e., the full set of individual nuisance parameters and *individual* diffusion coefficients were fitted for each run. Figure 6.7 shows the residuals $k - k_{\text{meas}}$ of all 6 measurements for the individual evaluations. Measurements (a1), (a2), (a3) and (a4) have residuals $k - k_{\text{meas}} < 0.2$ for most of the values and were accepted; measurements (b1) and (b2) were discarded due to large residuals $k - k_{\text{meas}} > 0.2$ with pronounced time dependencies. This time dependencies could, e.g., be due to more complicated temperature drifts of the optical setup than the linear dependence assumed. It is also important to notice that even the residuals of the accepted measurements exhibit remaining small time dependencies. Therefore, remaining small systematic errors have to be expected.

After identification of the acceptable measurements (a1)-(a4), we performed a simultaneous evaluation of the accepted measurements, i.e., the full set of individual nuisance parameters and a *common* diffusion coefficient were estimated. Figure 6.8 shows the residuals $k - k_{\text{meas}}$ of the accepted measurements for the simultaneous evaluation. The common diffusion coefficient of the simultaneous evaluation is a compromise for all suitable measurements. Therefore, the residuals are slightly larger for the simultaneous evaluation in comparison to the individual evaluation, but the increase in *RMSE* is small (between 0.01 and 0.03 fringe orders) indicating consistent experiments. Still, the overall accuracy is excellent. Based on the total number of approximately 120 interference fringes that occur during the diffusion process, the larger *RMSE* of 0.1 corresponds to a relative error of $0.03/120 = 0.025\,\%$.

Figure 6.9 shows the concentration-dependent diffusion coefficient from a single experiment plotted against the mole fraction of krypton x_{Kr} estimated from our experiments (lines) in comparison to our measurements with small initial composition differences and to literature values (symbols). The absolute values of the diffusion coefficients from the simultaneous evaluation and the individual evaluation agree very well. Furthermore, the slopes of the concentration dependencies obtained for all experiments with pure gases in this work agree well with the slopes from our measurements

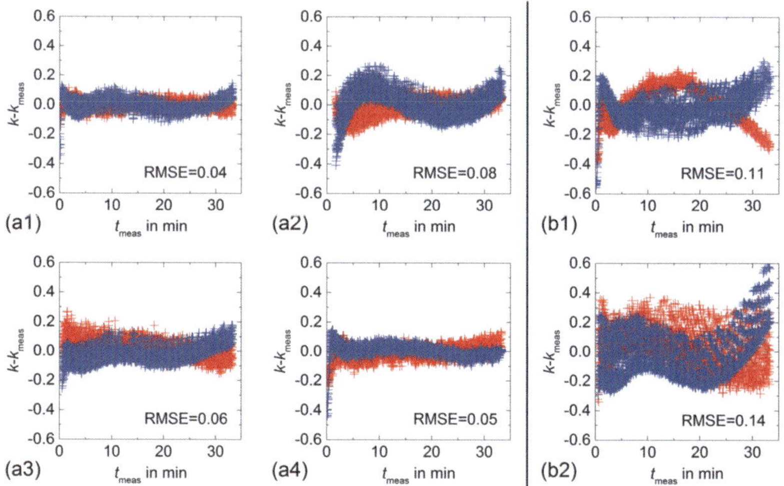

Figure 6.7: Residuals $k - k_{\mathrm{meas}}$ over time t_{meas} of all measurements when the parameter estimation procedure is applied to each measurement individually, i.e., the full set of individual nuisance parameters and *individual* diffusion coefficients are fitted. Red: residuals for the upper half-cell; blue: residuals for the lower half-cell. (a1), (a2), (a3), (a4): Residuals of accepted measurements, which exhibit acceptable residuals $k - k_{\mathrm{meas}} < 0.2$. (b1), (b2): Residuals of discarded experiments, which exhibit inacceptable residuals $k - k_{\mathrm{meas}} > 0.2$ and pronounced time dependencies. Root-mean square errors: $RMSE(\mathrm{a1}) = 0.04$, $RMSE(\mathrm{a2}) = 0.08$, $RMSE(\mathrm{a3}) = 0.06$, $RMSE(\mathrm{a4}) = 0.05$, $RMSE(\mathrm{b1}) = 0.11$, $RMSE(\mathrm{b2}) = 0.14$.

with small initial composition differences and with literature values. However, an offset of about 1.5 % is clearly visible. This offset is most likely due to remaining systematic errors.

Although the reason for the offset could not finally be clarified, we could exclude a number of possible error sources. For this purpose, we calculated the impact of potential erroneous parameters on the interference fringe orders: We used the diffusion model to calculate the deviations in the interference fringe orders when erroneous parameters are applied. Table 6.2 shows the values and associated uncertainties of the experimental setup and the system helium-krypton. We consider only parameters that

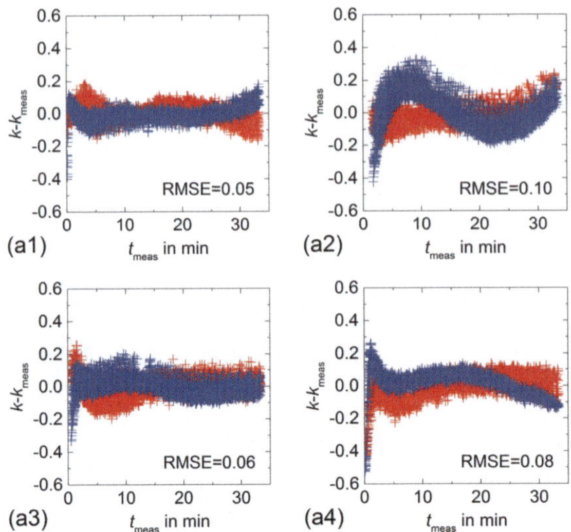

Figure 6.8: Residuals $k - k_{\mathrm{meas}}$ over time t_{meas} of all suitable experiments (a1), (a2), (a3), and (a4), respectively, when the parameter estimation procedure is applied to all suitable experiments simultaneously, i.e., the full set of individual nuisance parameters and a *common* diffusion coefficient is fitted. Red: residuals for the upper half-cell; blue: residuals for the lower half-cell. Root-mean square errors: $RMSE(\mathrm{a1}) = 0.05$, $RMSE(\mathrm{a2}) = 0.10$, $RMSE(\mathrm{a3}) = 0.06$, $RMSE(\mathrm{a4}) = 0.08$.

are not fitted in the parameter estimation procedure, since only non-fitted parameters can lead to systematic errors.

The measured pressures p_{meas} are not used in the data evaluation procedure. Instead, the initial pressures are fitted and the pressure is calculated from the diffusion model. Therefore, uncertainties in the measured pressures have no impact on the data evaluation. Uncertainties in temperature T, depth of the Loschmidt cell l, laser wavelength λ, refractivity virial coefficients $A_{\mathrm{R},i}$, pressure virial coefficients B_{ij} and C_{ijk}, and the initial compositions $x_{1,0}$ of the pure gases lead to uncertainties in the interference fringe orders which are at least one order of magnitude lower than the threshold of 0.2 that is required to retrieve the concentration dependence of the dif-

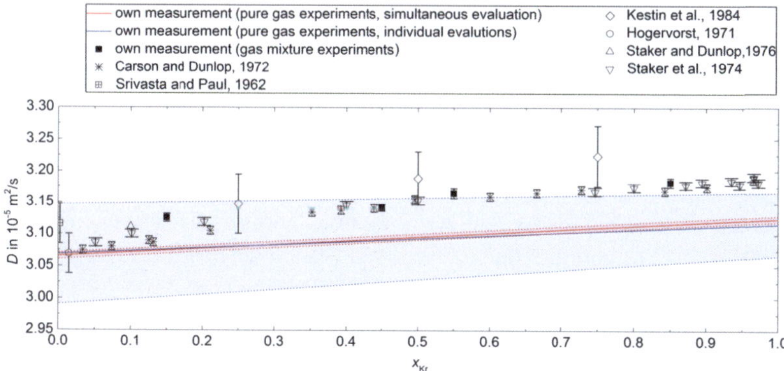

Figure 6.9: Experimental diffusion coefficients for the system helium-krypton at 293.15 K and 0.2 MPa as function of the mole fraction of krypton x_{Kr}. The error bars and shaded areas of our measurements represent 95 % confidence intervals. The literature data has been corrected to 293.15 K and 0.2 MPa with the relations $D \propto T^{11/6}$ (Marrero and Mason, 1972) and $D \propto 1/p$ (Cussler, 2007) if necessary. Numerical values of all concentration-dependent diffusion coefficients are provided in Appendix C.

fusion coefficient. Only uncertainties in the length L of the Loschmidt cell can lead to systematic errors in the interference fringe orders of 0.12 that are in the order of magnitude, but still below the threshold of 0.2.

The remaining thermal drifts in the temperature ϑ of the optical setup can lead to errors in the interference fringe orders of up to 0.8 (cf. Section 6.1.1). We accounted for the remaining thermal drifts with a linear drift parameter k_{drift}. Therefore, linear thermal drifts cannot lead to systematic errors in the interference fringe orders. We analyzed the movement of interference fringes after completion of the diffusion process to exclude experiments with large non-linear thermal drifts in the interference fringe orders (see above). Although the fitted linear drift parameters k_{drift} agree with the drifts of the interference fringe orders of the remaining experiments after completion of the diffusion process, the linear drift behavior is always just an approximation. Furthermore, it could be that an experiment had non-linear thermal drifts during the diffusion process (for approx. 30 min) and linear drifts only after completion of the diffusion process. Identification and fitting of such non-linear drifts is impossible with the current experimental setup and parameter estimation procedure. Therefore, remaining thermal drifts of the optical setup could be a reason for the offset in the concentration-dependent diffusion coefficient.

Overall, we could retrieve the slope of the concentration dependence of the diffusion coefficient from a single experiment. The reason for a remaining offset of 1.5 % in the absolute values of the diffusion coefficient could not finally be clarified. Further efforts should be spend to check the length L of the Loschmidt cell and on improving the stability of the optical setup, e.g. by application of interferometry stabilization techniques (MacQuigg, 1977; Neumann and Rose, 1967).

Table 6.2: Parameter values and uncertainties of the experimental setup and the system helium(2)-krypton(1) and consequential uncertainties in the interference fringe order.

Parameter	Symbol	Value & Uncertainty	Reference	Uncertainty Δk in interference fringe order
Pressure in MPa	p_{meas}	$0.2 \pm 0.06\,\% \cdot p_{\mathrm{meas}}$	Kugler (2015)	not used in data evaluation
Temperature in K	T	293.15 ± 0.035	Kugler (2015)	0.01
Depth of the Lohschmidt cell in mm	l	201 ± 0.05	Kugler (2015)	0.03
Laser wavelength in nm	λ	632.8 ± 0.1	Kugler (2015)	0.02
Refractivity virial coefficients in $\mathrm{cm}^3/\mathrm{mol}$	A_{R1}	6.362 ± 0.004	Achtermann et al. (1993)	0.01
	A_{R2}	0.5213 ± 0.004	Achtermann et al. (1993)	
2nd pressure virial coefficient in $\mathrm{cm}^3/\mathrm{mol}$	B_{11}	-52.666 ± 0.3	Kugler (2015)	0.01
	B_{22}	11.847 ± 0.3	Kugler (2015)	
	B_{12}	21.517 ± 0.3	Kugler (2015)	
3rd pressure virial coefficient in $\mathrm{cm}^3/\mathrm{mol}$	C_{111}	2300 ± 300	Kugler (2015)	
	C_{222}	1109 ± 300	Kugler (2015)	
	C_{112}	$(C_{111}^2 C_{222})^{1/3}$	Kugler (2015)	
	C_{122}	$(C_{111} C_{222}^2)^{1/3}$	Kugler (2015)	
Gas purity	$x_{1,0}$	$1 - 0.00001$	Linde®	0.002
Length of the Loschmidt cell in mm	L	397.14 ± 0.04	Kugler (2015)	0.12
Temperature of the optical setup in °C	ϑ	20 ± 0.1	this work	0.8

6.1.6 Conclusions

The measurement of concentration-dependent gaseous diffusion coefficients from a single experiment with pure gases has been a long-time challenge for scientists. For this task, both highly accurate measurements and an appropriate diffusion model are required. In the past, the measurement of concentration-dependent gaseous diffusion coefficients with a Loschmidt cell combined with interferometry showed discrepancies between experiment and literature values (Baranski, 2002; Buttig et al., 2011; Kugler et al., 2013, 2015a,b; Kullnick, 2001). To resolve the discrepancies, we performed a systematic experimental analysis. We developed an improved experimental setup and an improved data evaluation procedure. The data evaluation employs a consistent diffusion model including real gas effects and the concentration dependence of the diffusion coefficient. We found that it is crucial to account for uncertainties (1) in the initial conditions, (2) in the thermal stability of the optical setup, and (3) in the camera calibration. The experimental setup has been improved by (1) introducing initial interference fringes instead of an initial white zero field, (2) controlling the temperature of the optical setup, and (3) installation of a calibration reference. The data evaluation procedure has been improved by introducing the following nuisance parameters into the diffusion model: (1) the initial interference fringe order, (2) a linear drift correction accounting for remaining small temperature drifts of the optical setup, and (3) a correction for offsets in the measurement locations.

The improved measurement setup and evaluation procedure have been validated with diffusion measurements of the system helium-krypton. We performed multiple experiments with gas mixtures of various initial compositions in the half-cells of the Loschmidt cell. The concentration dependence of the diffusion coefficient was successfully retrieved. The results agree with literature data. The excellent quality of fit emphasizes the high confidence in the results.

Subsequently, we performed diffusion measurements with pure helium and krypton. The model describes the experimental fringe data with deviations of less than 0.2 interference fringe orders, which corresponds to a relative deviation of 0.17 % indicating high quality of both the experimental data and the employed model. The concentration dependence of the diffusion coefficient could therefore be successfully retrieved from a single experimental run. The slope of the concentration-dependence agrees well with our validated results from multiple experiments with gas mixtures. A small offset of 1.5 % in the diffusion coefficient remains but could potentially be further reduced by improving the thermal stability of the optical setup. Thus, the presented combined approach of experiment and data evaluation allows for the efficient and fast characterization of diffusion in gases.

6.2 Diffusion in liquids: Microfluidic experiments

Diffusion experiments with liquids are burdened with long measurement times and large sample consumption (cf. Chapter 2). In particular, multicomponent diffusion measurements are associated with large experimental effort, because multiple experiments are typically required to determine multicomponent diffusion coefficients.

Microfluidics is particularly suited to overcome the main limitations in the measurement of liquid diffusion coefficients: Small dimensions result in short measurement times ($<1\,\mathrm{h}$), since the diffusion time is proportional to the squared diffusion distance (cf. Section 2.1.1). Furthermore, microfluidics allows for reduced sample volumes due to the small channel volumes ($\approx 10\,\mu\mathrm{L}$) (Kamholz et al., 2001; Hatch et al., 2001). In addition, the steady-state microfluidic diffusion process allows for easy repetitions of experiments.

The potential and the feasibility of microfluidic diffusion measurements have previously been demonstrated for binary mixtures (Kamholz et al., 2001; Pappaert et al., 2005; Häusler et al., 2012; Costin et al., 2013; Werts et al., 2012; Ryckeboer et al., 2013; Lefortier et al., 2012; Munson et al., 2005). It has already been suggested that microfluidics offers the potential for the measurement of multicomponent diffusion coefficients (Häusler et al., 2012; Salmon et al., 2005; Wang et al., 2005). However, to the best of the authors' knowledge, microfluidic diffusion measurements of multicomponent systems have not been accomplished yet.

To investigate diffusion in multicomponent systems, Raman spectroscopy is a particularly suitable analytical tool as shown for macroscopic experiments (Bardow et al., 2006). Raman spectroscopy allows for the simultaneous quantification of all components with high accuracy and high spatial resolution (Pelletier, 2003; Salmon et al., 2005). Thereby, it is possible to determine multicomponent diffusion coefficients from a single experiment (Bardow et al., 2006).

In this section, we present a multicomponent diffusion experiment combining a microfluidic device with Raman microspectroscopy. A model-based experimental approach is applied for an optimal design of the experimental setup and the consistent evaluation of measurement data. The new setup allows for the measurement of multicomponent liquid diffusion coefficients from a single experiment. Thereby, significant savings in experimental effort and sample consumption are achieved.

6.2.1 Experimental setup and procedure

The experimental setup is based on Raman microspectroscopy in an H-cell microchannel (cf. Figure 6.10). Inside the H-cell, two solutions of different compositions are contacted and the compositions equilibrate through diffusion perpendicular to the flow direction.

The microfluidic chip (Micronit, Netherlands) consists of two bonded borosilicate glass plates. The H-cell is a custom-made channel in form of a spiral of length $L = 1\,\mathrm{m}$, width $W = 400\,\mu\mathrm{m}$, and height $H = 40\,\mu\mathrm{m}$ (cf. Figure 6.10(a)). The temperature of the microfluidic chip is controlled by a thermostat (CC-K15 Pilot One, HUBER SE, Germany) with a precision of $\pm 0.5\,\mathrm{K}$.

The microfluidic glass chip provides *in situ* access to the diffusion process by non-invasive Raman microspectroscopy. Four observation points s_1 to s_4 are chosen (illustrated in Figure 6.10(a)) to observe the diffusion process at several contact times (i.e., various Fourier numbers Fo). At each observation point, Raman spectra are acquired every $2.5\,\mu\mathrm{m}$ along the channel cross-section y (illustrated in Figure 6.10(c)). These Raman spectra are used to obtain spatially resolved composition profiles.

6.2.2 Data evaluation

The experimental data consists of a set Raman spectra recorded at each spatial resolution point y and each observation point s. The Raman spectra are analyzed using the indirect hard modeling (IHM) method (Alsmeyer et al., 2004). In the IHM analysis, every pure component spectrum is modeled by a sum of peak-shaped pseudo-Voigt functions (the so-called component model). The Raman spectrum of the borosilicate glass is an additional component model used as background. All component models of the system are weighted and added to a mixture model.

To calculate the mole fractions from a mixture spectrum, the residuals between the spectrum and the mixture model are minimized in a least squares procedure. From the components in the mixture model, integrated intensities of the pure components

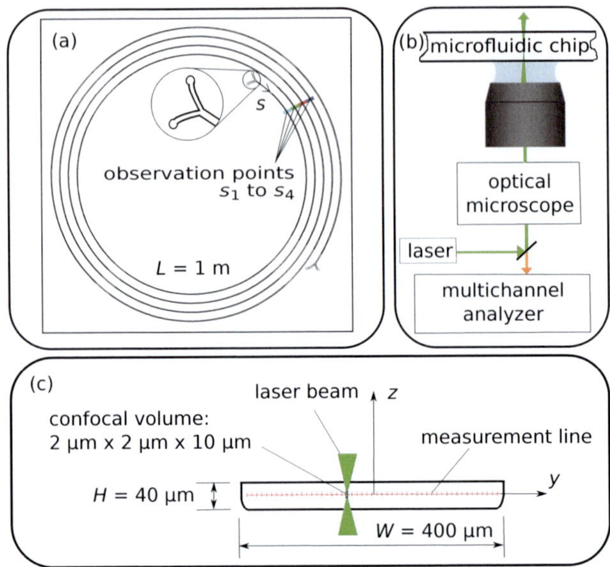

Figure 6.10: (a) Top view of microfluidic chip with microchannel in form of a spiral H-cell with observation points s_1 (–), s_2 (–), s_3 (–), and s_4 (–). Zoom to inlet of H-cell. (b) Optical setup consisting of laser, microfluidic chip, optical microscope, and multichannel analyzer. (c) Channel cross-section with dimensions, coordinate system, laser beam, confocal volume, and measurement line (+).

A_i are obtained for each spectrum, which are converted to mole fractions

$$x_{i,\mathrm{meas}} = \frac{A_i}{\sum\limits_{j=1}^{n_c} A_j k_{i,j}} \tag{6.19}$$

using calibration factors $k_{i,j}$. The calibration factors $k_{i,j}$ are obtained from the analysis of spectra from mixtures of known composition. The spectra to determine the calibration factors $k_{i,j}$ are also obtained in the H-cell.

Convection-diffusion process model

The convection-diffusion process model is based on a model developed by Häusler et al. (2012) for binary mixtures. We assume a no-slip, laminar flow in a duct with a fully developed and gravimetrically stable flow field. This assumption is based on the fact that the channel is flat ($W \gg H$, cf. Figure 6.10(c)) and that the magnitude of the Reynolds number Re is of order 1 (Lin et al., 2012).

The diffusion process is continuous and steady-state. Convection dominates the mass transport in length-direction s (Péclet number $Pe > 100$), while diffusion dominates the mass transport in the y-z cross-sectional plane. Furthermore, we average the concentrations over the z-direction because concentrations are smoothed out by diffusion much faster in z-direction than in y-direction due to the high aspect ratio ($W \gg H$).

The channel is fed with solutions composed of mole fractions $x_{\mathrm{l},i}$ and $x_{\mathrm{r},i}$ of component i for the left (l) and right (r) channel inlet, respectively, i.e.,

$$x_i(s=0,y) = \begin{cases} x_{\mathrm{l},i} & \text{for } y < y_0 \\ x_{\mathrm{r},i} & \text{for } y \geq y_0 \end{cases}, \qquad i = 1, ..., n_{\mathrm{c}} - 1\,. \tag{6.20}$$

Here, y_0 denotes the position where the two channel inlets join. The difference in mole fractions $\Delta x_i = |x_{\mathrm{l},i} - x_{\mathrm{r},i}|$ between the left and right channel inlet is chosen preferably small. Thereby, constant viscosities and negligible excess volume effects can be assumed ($\boldsymbol{V}^{\mathrm{E}} = 0$). The diffusion coefficient matrix $\boldsymbol{D}$ is also averaged over the measured mole fraction range Δx_i. A constant diffusion coefficient matrix $\boldsymbol{D}$ can be assumed for the small Δx_i employed here.

Using Fick's law to describe the diffusive fluxes (Equation 2.5), the multicomponent convection-diffusion model then reads:

$$v(y) \cdot \frac{\partial c_i(s,y)}{\partial s} = \sum_{j=1}^{n_{\mathrm{c}}-1} D_{i,j}^{V} \frac{\partial^2 c_j(s,y)}{\partial y^2} \quad \text{for } i = 1, ..., n_{\mathrm{c}} - 1, \tag{6.21}$$

$$x_i(s,y) = \frac{c_i(s,y)}{c_{\mathrm{t}}(s,y)} \quad \text{for } i = 1, ..., n_{\mathrm{c}}, \quad c_{\mathrm{t}}(s,y) = \sum_{i=1}^{n_{\mathrm{c}}} c_i(s,y) \quad \text{and}$$

$$\frac{1}{c_{\mathrm{t}}(s,y)} = \sum_{i=1}^{n_{\mathrm{c}}} x_i(s,y) \cdot V_i^0\,.$$

Here, V_i^0 is the molar volume of pure component i.

The z-averaged velocity profile $v(y)$ for a laminar, no-slip flow in a rectangular duct is given by (Rohsenow et al., 1998)

$$v(y) = \left(\frac{m+1}{m}\right)\left[1 - \left(\frac{y}{(W/2)}\right)^m\right] \cdot \bar{v} \tag{6.22}$$
$$\text{with } m = 1.7 + 0.5\left(\frac{H}{W}\right)^{-1.4} .$$

Here, $\bar{v}$ denotes the average velocity in the channel given by $\bar{v} = \dot{V}/(H \cdot W)$. $\dot{V}$ is the total volume flow in the channel.

In addition to the inlet conditions 6.20, the convection-diffusion model is complemented by the two no-flux boundary conditions at the channel walls:

$$\frac{\partial c_i(s, -W/2)}{\partial y} = \frac{\partial c_i(s, W/2)}{\partial y} = 0, \quad i = 1, ..., n_c - 1. \tag{6.23}$$

Parameter estimation

The mole fractions $x_{i,\text{meas}}$ obtained from the recorded Raman spectra are used as measurement data for the parameter estimation. The standard deviation $\sigma(x_i)$ of the mole fractions is assumed to be constant and to be the same for all components of a mixture. Therefore, unweighted least squares are used in the parameter estimation. The parameters are obtained from the least-squares minimization problem (cf. Equation 3.30):

$$\Phi_{\text{WLS}} = \min_{\boldsymbol{\theta}} \sum_{\mu=1}^{N_{\text{exp}}} \sum_{i=1}^{N_c - 1} \sum_{o=1}^{N_s} \sum_{p=1}^{N_R(o)} \left(x_{\text{calc},\mu,i,o,p} - x_{\text{meas},\mu,i,o,p}\right)^2 . \tag{6.24}$$

Here, N_{exp} denotes the number of experiments, N_c the number of components, N_s the number of observation points, and $N_R(o)$ the number of Raman spectra along the measurement line at observation point s_o. x_{calc} are the mole fractions calculated from the convection-diffusion model (cf. Equations 6.20-6.23. The parameters $\boldsymbol{\theta}$ to be estimated are $\boldsymbol{\theta} = \left[\boldsymbol{D}^V, \boldsymbol{\Delta s}, \boldsymbol{y}_0, \boldsymbol{x}_l, \boldsymbol{x}_r\right]$ and explained in more detail in the following. All estimated parameters besides the diffusion coefficient matrix $\boldsymbol{D}^V$ are estimated for each experiment individually. All estimated parameters besides the diffusion coefficient matrix $\boldsymbol{D}^V$ are nuisance parameters which are not of immediate interest but have to be accounted for to get statistically sound results:

- Δs is an offset of the observation point s_1 to the inlet. One Δs is adjusted for each experiment; it is typically in the range of 0 µm to 2 µm. The distances between the observation points s_o are fixed.
- y_0 adjusts the position where the two channel inlets join. One y_0 is adjusted for each experiment; it is typically in the range of −5 µm to 5 µm.
- The mole fractions $\boldsymbol{x}_\mathrm{l}$ and $\boldsymbol{x}_\mathrm{r}$ at the left and right inlet, respectively, are fitted once for each experiment to account for small uncertainties in the mixture compositions.

The diffusion coefficient $\boldsymbol{D}^V$ is positive definite according to the thermodynamics of irreversible processes Taylor and Krishna (1993). The corresponding constraints are directly incorporated into the fitting procedure by a reparameterization proposed by Bardow et al. (2006) (cf. Appendix D.2).

6.2.3 Results and discussion

To validate the microfluidic experimental setup and model-based experimental analysis, we performed diffusion measurements with the binary system cyclohexane + toluene. Subsequently, we performed measurements with the ternary system 1-propanol + 1 chlorobutane + heptane at 293.15 K to demonstrate the significant savings in experimental efforts when using Raman microspectroscopy.

Binary: cyclohexane + toluene

For cyclohexane + toluene, the average composition of the mixture was chosen to be same as used by Sanni et al. (1971): $\bar{x}_\mathrm{cyclohexane} = 60\mathrm{mol}\,\%$, $\bar{x}_\mathrm{toluene} = 40\mathrm{mol}\,\%$. The mole fraction difference between the inlet compositions was chosen to be $\Delta x_i = 3\,\mathrm{mol}\,\%$. On the one hand, this Δx_i is large enough to distinguish the mixture spectra. On the other hand, this Δx_i is small enough to assume a constant diffusion coefficient.

The volume flow $\dot{V}$ is adjusted to ensure Fourier numbers growing from $Fo = 0.0025$ to $Fo = 0.1$ at the observation points s_1 to s_4. Thus, the optimum Fourier number as identified by OED techniques are included in the experiment (cf. Chapter 5). Within

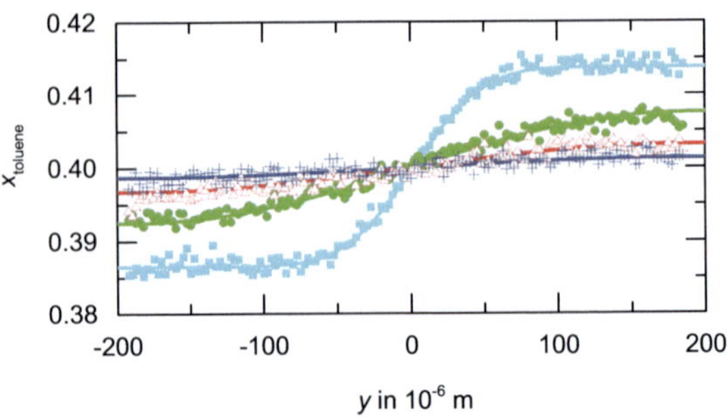

Figure 6.11: Measured mole fraction profiles (symbols) and fitted mole fraction profiles (lines) of toluene in the system cyclohexane + toluene along measurement line y of the channel cross-sections at observation points $s_1 = 0.0143\,\mathrm{m}$ ($\bar{t} = s/\bar{v} = 0.25\,\mathrm{s}$, ■,), $s_2 = 0.225\,\mathrm{m}$ ($\bar{t} = 7.41\,\mathrm{s}$, •,), $s_3 = 0.452\,\mathrm{m}$ ($\bar{t} = 14.89\,\mathrm{s}$, △,), and $s_4 = 0.694\,\mathrm{m}$ ($\bar{t} = 22.86\,\mathrm{s}$, +,).

the chosen range of Fourier numbers, there is considerable progress of diffusional mixing across the whole channel width while the concentration change over the channel width is still observable.

The diffusion experiment has been repeated 4 times. Figure 6.11 shows an example of the measured and calculated mole fraction profiles at observation points s_1 to s_4. As expected, the mole fraction gradient decreases with consecutive observation points s_o. Measured and calculated mole fractions are in excellent agreement. Considering all experiments, the root mean square error is $RMSE = 0.088\,\mathrm{mol}\,\%$.

Table 6.3 gives values of the average of the diffusion coefficient and its standard deviation obtained from our experiments and from Sanni et al. (1971). The diffusion coefficients agree excellently within measurement uncertainty: The relative deviation between the diffusion coefficients is 0.45 %. Thus, the experimental setup and data evaluation enable the successful determination of diffusion coefficients in liquids.

Ternary: 1-propanol + 1-chlorobutane + heptane

The full potential of Raman microspectroscopy can be exploited with multicomponent systems: Since Raman spectroscopy allows for the simultaneous quantification of all

Table 6.3: Measured diffusion coefficient for the binary system cyclohexane + toluene at $x_{\text{cyclohexane}} = 60\,\text{mol}\,\%$ and 25 °C from microfluidic experiments in this work with standard deviation $\sigma(\boldsymbol{D}^V)$ from four repeated experiments in comparison to results from Sanni et al. (1971).

	$\boldsymbol{D}^V$	$\sigma(\boldsymbol{D}^V)$
	in $10^{-9}\,\text{m}^2\,\text{s}^{-1}$	
this work	1.759	0.042
Sanni et al. (1971)	1.767	0.018 to 0.035

components, a single experiment should be sufficient to determine the multicomponent diffusion coefficient matrix. However, OED calculations from Bardow et al. (2006) suggest that the accuracy of the multicomponent diffusion coefficient matrix can be significantly improved if a second, independent experiment is added that has the same average composition, but differs maximally in the compositions of the inlet solutions. To test both options of a single experiment and of two independent experiments, we performed two series of independent experiments: series A and series B. The diffusion experiments were performed with the system 1-propanol + 1-chlorobutane + heptane. The mixture compositions were chosen to be the same as used by Käshammer et al. (1994). The experiment of series A was repeated 4 times, the experiment of series B was repeated 5 times.

The diffusion coefficient matrix $\boldsymbol{D}^V$ was estimated by two procedures:

- In the first procedure, the diffusion coefficient matrix was determined for every individual experiment ($N_{\text{exp}} = 1$) from series A and series B, e.g., experiment A.1.
- In the second procedure, the diffusion coefficient matrix was determined for all combinations of one experiment from series A with one experiment of series B ($N_{\text{exp}} = 2$), e.g., A.1 with B.3.

As in the binary case, measured and calculated mole fractions are in excellent agreement; the *RMSE* is less than 0.4 mol % (an example of measured and calculated mole fraction profiles is given in Appendix D.3). The agreement between model and experiment retained the same excellent quality independently whether the diffusion coefficient was determined from individual experiments or two independent experiments (cf. Appendix D.3). Hence, the convection-diffusion model is suitable to describe the

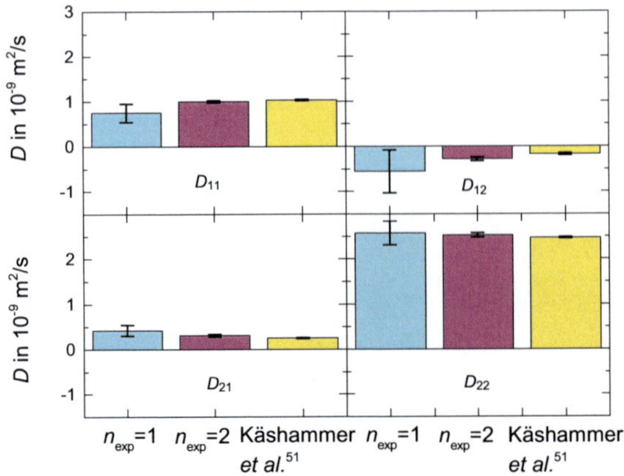

Figure 6.12: Diffusion coefficient matrix $\boldsymbol{D}^V$ with standard deviations of 1-propanol + 1-chlorobutane + heptane at $x_1 = x_2 = 33\,\text{mol}\,\%$ and $T = 25\,°\text{C}$ from this study (■, ■) and from Käshammer et al. (1994) (■). For $n_{\text{exp}} = 1$ (■), $\boldsymbol{D}^V$ is calculated from a single microfluidic Raman experiment. For $n_{\text{exp}} = 2$ (■), $\boldsymbol{D}^V$ is calculated from two independent experiments. Numerical values are given in Table D.4.

mass transport for multicomponent systems.

Results for the **estimation of a ternary diffusion coefficient from a single experiment** are shown in the left bars in the matrix of Figure 6.12 ($N_{\text{exp}} = 1$). Numerical results are given in Appendix D.3. Although only a single experiment was performed, reasonable precision and accuracy of the diffusion coefficient $\boldsymbol{D}^V$ were achieved: The relative errors between our diffusion coefficients (D_{ij}^V) and literature values is less than 33 %. The relative standard deviations of our diffusion coefficients is less than 16 %. The averaged diffusion coefficient matrix agrees with literature data (Käshammer et al., 1994) within the measurement uncertainty. Thus, a single experiment is sufficient to obtain the multicomponent diffusion coefficient matix with reasonable accuracy

The **estimation of a ternary diffusion coefficient from two independent experiments** should increase the accuracy significantly. The ternary diffusion coefficient matrix $\boldsymbol{D}^V$ from combinations of two experiments (one each from series A and series B) is shown in the central bars in the matrix of Figure 6.12 ($N_{\text{exp}} = 2$).

The standard deviation of the diffusion coefficients D_{ij}^{V} decreases by a factor of 10 in comparison to the estimation from single experiments. The relative standard deviations of our main diffusion coefficients is less than 2.5 %; the relative standard deviations of our cross diffusion coefficients is less than 16 %. The main diffusion coefficients and the determinant of the diffusion coefficient matrix agree with the literature data within the measurement uncertainty. Thus, 2 independent experiments are sufficient to determine multicomponent diffusion coefficients with high accuracy.

6.2.4 Conclusions

An experimental setup combining microfluidics with Raman microspectroscopy has been developed that enables the measurement of multicomponent diffusion coefficients in liquids. The experimental setup was validated with binary measurements of the system cyclohexane + toluene. Subsequently, we performed diffusion measurements with the ternary system 1-propanol + 1-chlorobutane + heptane. To the best of the authors' knowledge, these ternary diffusion measurements have been the first multicomponent diffusion measurements in a microfluidic device.

The experimental results show that a single experiment suffices to determine ternary diffusion coefficients with a reasonable accuracy of $\sigma(\boldsymbol{D}^V) = 33\,\%$. By combining two independent experiments, the accuracy could be significantly increased to $\sigma(\boldsymbol{D}^V) = 3.1\,\%$.

Thus, the new experimental setup enables significant savings in experimental effort and sample consumption. In addition, the experiment duration of less than 1 h is even competitive to the fastest classical diffusion experiments. Meanwhile, the experimental setup and data evaluation procedure has been successfully applied to quaternary mixtures (Peters et al., 2019).

Chapter 7

Optimal experimental design of liquid-liquid equilibria experiments

Contribution of the author: Co-author in writing the draft and final text, modeling, analysis.

Since predictive methods are still not sufficiently accurate, experiments are indispensable for the characterization of LLE (cf. Chapter 2). Typically, LLE experiments are performed to parametrize g^{E}-models which in turn are used for LLE predictions in process simulations. However, LLE experiments are burdened with long equilibration times and large sample consumption (cf. Chapter 2). Therefore, a reduction of the *number* of experiments would be desirable if the accuracy of the g^{E} model parameters and the LLE predictions would be retained. This goal can be achieved by use of optimal experimental design techniques (cf. Chapter 3).

In this chapter, optimal experimental designs for ternary LLE measurements are identified. An LLE experiment essentially consists of the preparation of an overall composition of components, $\boldsymbol{z}$, which separates into two phases ' and " with phase compositions $\boldsymbol{x}'$ and $\boldsymbol{x}''$, which are analyzed subsequently (cf. Section 2.2.1). Hence, the aim of OED for LLE is to determine optimal overall compositions $\boldsymbol{z}$ to be used in the experiments. Performing the optimal experiments, the number of experiments can be reduced without losing accuracy of the LLE predictions.

7.1 Method

The LLE experiment is described by the process model introduced in Section 2.2.2. The activity coefficients $\gamma_i(\boldsymbol{\theta}, \boldsymbol{x})$ are calculated using NRTL (Renon and Prausnitz, 1968) and UNIQUAC (Abrams and Prausnitz, 1975), respectively. The interaction parameters τ_{ij} of the g^{E}-models are the model parameters $\boldsymbol{\theta}$ we want to obtain from the experimental procedure. The Fisher information matrix is obtained from Equation 3.8. Thus, we need the derivatives of the model equations $\boldsymbol{g}(\boldsymbol{w}, \boldsymbol{z}, \boldsymbol{\theta})$ with respect to the parameters. For nonlinear models, these derivatives depend on the values of the parameters. Hence, we need an initial estimate of these parameters, $\hat{\boldsymbol{\theta}}$ (cf. Section 3.2.1). This estimate could result from predictive models, such as UNIFAC (Fredenslund et al., 1975) or COSMO-RS (Klamt, 1995). For the analysis in this work, we consider systems with known LLE and known parameters $\hat{\boldsymbol{\theta}}$. If parameters are to be fitted from LLE experiments, robust estimation methods should be applied that ensure global unique solutions (Mitsos et al., 2009).

In order to simplify the representation of the overall composition $\boldsymbol{z}$, only overall compositions in the middle of the tie-lines are considered in the following. This allows to replace the vectorial quantity $\boldsymbol{z}$ by a linear scalar measure $\alpha \in [0, 1]$. α is 0 at the binary subsystem, and α is 1 at the critical point in type I LLE and at the second binary subsystem in type II LLE, respectively (see figure 7.1). The scaling of α is not necessary; one could optimize the composition $\boldsymbol{z}$ directly. Since g^{E}-models cannot describe LLE with sufficient accuracy near the critical point Gmehling et al. (2012), we consider only overall compositions in the range $\alpha \in [0, 0.8]$. For numerical calculations, the design space α was discretized into 100 commensurate sections.

The objective function for the optimal design is the minimization of the variance of the predictions. Therefore, a G-optimal design has to be determined. According to the general equivalence theorem (Equation 3.21), a continuous G-optimal design is equal to a continuous D-optimal design. Thus, the optimal experimental design minimizing the variance in the LLE predictions also minimizes the variance in the g^{E} model parameters.

However, for practical applications, *exact* optimal designs with a finite number of measurements have to be determined instead of continuous designs. For this purpose, we consider LLE experiments with a total number of 10 measurements. In a first step, we calculate the continuous D/G-optimal design $\boldsymbol{d}^{*}_{\mathrm{D/G-opt}}$. Subsequently, the exact D/G-optimal design $\boldsymbol{d}^{(10)}_{\mathrm{D/G-opt}}$ is built by rounding the design weights of the continuous design $\boldsymbol{d}^{*}_{\mathrm{D/G-opt}}$.

To evaluate the quality of the optimal designs, conventional, intuitive designs, $\boldsymbol{d}^{(10)}_{\mathrm{Conv1}}$ and $\boldsymbol{d}^{(10)}_{\mathrm{Conv2}}$, are introduced as a benchmark. Each conventional design contains also a total number of 10 experiments. The conventional designs were chosen to have equidistantly distributed experiments: The first conventional design $\boldsymbol{d}^{(10)}_{\mathrm{Conv1}}$ contains as many distinct experiments as the optimal design, whereas the second conventional design $\boldsymbol{d}^{(10)}_{\mathrm{Conv2}}$ always contains 10 equally distributed, distinct experiments (i.e. without repetition of any measurement).

The measurements of the LLE phase compositions x_i' and x_i'' of each component i are assumed to be independent of each other. Therefore, the measurement variance matrix $\boldsymbol{V_w}$ is assumed to be diagonal with equal diagonal entries:

$$\boldsymbol{V}_w = \boldsymbol{I} \cdot \sigma_x^2. \tag{7.1}$$

The value of σ_x^2 does not affect the optimal design: Any diagonal measurement variance matrix $\boldsymbol{V}_w$ with equal diagonal entries will give the same results, since our objective function (Equation 3.18, G-optimality) is scaled by the measurement variance.

7.2 Results

We consider LLE of type I with a single miscibility gap and LLE of type II with two miscibility gaps. For type I, the system water(1)-acetone(2)-toluene(3) is used as a representative example. For type II, the system 3-methyltetrahydrofurane(1)-di-n-butylether(2) - water(3) is used. In Section 7.2.1, the optimal as well as the conventional designs for type I LLE described by UNIQUAC are presented and discussed in detail. In Section 7.2.2, we give a brief summary of the optimal designs for a type I LLE described by NRTL, and the optimal designs for a type II LLE described by UNIQUAC and NRTL. Details of the results can be found in Appendix F.

7.2.1 LLE of type I with UNIQUAC

For the type I LLE water(1)-acetone(2)-toluene(3), OED yields a continuous D/G-optimal design $\boldsymbol{d}^{*}_{\mathrm{D/G-opt}}$ which concentrates on three distinct positions:

$$\boldsymbol{d}^{*}_{\mathrm{D/G-opt}} = \left\{ \begin{array}{ccc} 0.22 & 0.57 & 0.8 \\ 0.24 & 0.30 & 0.46 \end{array} \right\}. \tag{7.2}$$

This means that 24% of the experimental effort is spent at the overall composition $\alpha = 0.22$, 30% of the experimental effort is spent at the overall composition $\alpha = 0.57$, and 46% of the experimental effort is spent at the overall composition $\alpha = 0.8$ (near the critical point).

An approximation of the exact optimal design $\boldsymbol{d}^{(10)}_{\mathrm{D/G-opt}}$ is built by rounding the design weights:

$$\boldsymbol{d}^{(10)}_{\mathrm{D/G-opt}} = \left\{ \begin{array}{ccc} 0.22 & 0.57 & 0.8 \\ 0.2 & 0.3 & 0.5 \end{array} \right\}. \tag{7.3}$$

Thus, two experiments are executed at the overall composition $\alpha = 0.22$, three experiments are executed at the overall composition $\alpha = 0.57$, and five experiments are executed at the overall composition $\alpha = 0.8$. Figure 7.1 (left) shows the positions of the experiments belonging to the optimal design $\boldsymbol{d}^{*}_{\mathrm{D/G-opt}}$ (respectively $\boldsymbol{d}^{(10)}_{\mathrm{D/G-opt}}$) in a ternary phase diagram, i.e. where experiments should be executed to minimize the variance of the predictions made by UNIQUAC.

To benchmark the performance of the computed optimal design, conventional, intuitive designs are used. The first conventional design $\boldsymbol{d}^{(10)}_{\mathrm{Conv1}}$ contains as many distinct experiments as the optimal design (three in this case), which are evenly distributed in the design space $\alpha \in [0, 0.8]$:

$$\boldsymbol{d}^{(10)}_{\mathrm{Conv1}} = \left\{ \begin{array}{ccc} 0 & 0.4 & 0.8 \\ 0.2 & 0.3 & 0.5 \end{array} \right\}. \tag{7.4}$$

The phase diagram in Figure 7.1 (center) shows the positions of the experiments belonging to the conventional designs $\boldsymbol{d}^{(10)}_{\text{Conv1}}$.

Since a conventional design consisting of only three distinct experiments is uncommon in practice, we consider another conventional design $\boldsymbol{d}^{(10)}_{\text{Conv2}}$ as a second benchmark. This second conventional design consists of 10 distinct experiments which are evenly distributed in the design space $\alpha \in [0, 0.8]$:

$$\boldsymbol{d}^{(10)}_{\text{Conv2}} = \left\{ \begin{array}{cccccccccc} 0 & 0.09 & 0.18 & 0.27 & 0.36 & 0.44 & 0.53 & 0.62 & 0.71 & 0.8 \\ 0.1 & 0.1 & 0.1 & 0.1 & 0.1 & 0.1 & 0.1 & 0.1 & 0.1 & 0.1 \end{array} \right\}. \tag{7.5}$$

Figure 7.1 (right) shows the positions of the experiments belonging to the second conventional designs $\boldsymbol{d}^{(10)}_{\text{Conv2}}$.

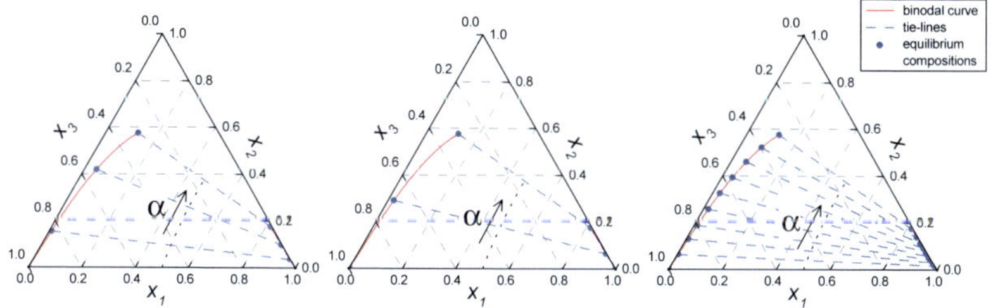

Figure 7.1: Positions of the optimally designed experiments and the equally distributed conventional experiments in the ternary phase diagrams.
Left: $\alpha(\boldsymbol{d}^{(10)}_{\text{opt}})$ (cf. Equation 7.3); center: $\alpha(\boldsymbol{d}^{(10)}_{\text{Conv1}})$, using three distinct experiments (cf. Equation 7.4); right: $\alpha(\boldsymbol{d}^{(10)}_{\text{Conv2}})$, using 10 distinct experiments (cf. Equation 7.5).

To evaluate the prediction accuracy that is expected from the optimal and conventional LLE experiments, we compare the standardized variances of the LLE predictions $\tilde{\sigma}_{\text{Pred}}$ (cf. Equation 3.19). Figure 7.2 shows the predicted standardized variances $\tilde{\sigma}_{\text{Pred}}$ plotted as a function of α. For the continuous optimal design $\boldsymbol{d}^{*}_{\text{opt}}$, we find that

$$\tilde{\sigma}_{\text{Pred}}\left(\boldsymbol{d}^{*}_{\text{opt}}\right) \leq 6, \tag{7.6}$$

where 6 is the number of parameters in UNIQUAC for a ternary system. Thus, the computed optimal design is optimal according to the general equivalence theorem (cf. Equation 3.22). The values of the predicted standardized variances of the exact optimal design, $\tilde{\sigma}_{\text{Pred}}\left(\boldsymbol{d}_{\text{opt}}^{(10)}\right)$, agree well with the values of the predicted standardized variances of the continuous optimal design, $\tilde{\sigma}_{\text{Pred}}\left(\boldsymbol{d}_{\text{opt}}^{*}\right)$. Hence, the exact design $\boldsymbol{d}_{\text{opt}}^{(10)}$ is a good approximation of the continuous design $\boldsymbol{d}_{\text{opt}}^{*}$. The prediction variances of the continuous as well as of the exact optimal design are evenly distributed and almost independent of α. Therefore, we expect an equal quality of the predictions along the binodal curve when optimal designs are performed.

In contrast, the prediction variance of the first conventional design $\boldsymbol{d}_{\text{Conv1}}^{(10)}$ is the largest for almost any value of α, especially at the large local maxima in between the experimental supporting points: The maximum of $\tilde{\sigma}_{\text{Pred}}\left(\boldsymbol{d}_{\text{Conv1}}^{(10)}\right)$ is more than three times higher than the maximum of $\tilde{\sigma}_{\text{Pred}}\left(\boldsymbol{d}_{\text{opt}}^{(10)}\right)$. Thus, the total number of experiments in the first conventional design $\boldsymbol{d}_{\text{Conv1}}^{(10)}$ has to be increased threefold to achieve a comparable accuracy in the predictions as in the optimal design $\boldsymbol{d}_{\text{opt}}^{(10)}$. It can therefore be concluded that the information gain on the shape of the binodal curve is low when executing experiments at the binary subsystem.

Regarding the prediction variance of the second conventional design $\boldsymbol{d}_{\text{Conv2}}^{(10)}$, an even distribution with low values over a large range of α can be observed. Only at high values of α, a drastic increase occurs. This region corresponds to the high curvature region of the binodal curve towards the critical point. This increase can be explained by the distribution of the experimental effort on the experimental positions: In the optimal designs, more effort is spent on the region near the critical point than in the second conventional design. Hence, a lack of sampling in the region near the critical point leads to insufficient information on the degree of curvature of the binodal curve.

7.2.2 LLE of type I system with NRTL and LLE of type II system with NRTL and UNIQUAC

In the following, we give a brief summary of the OED for a type I LLE described by NRTL, and the OEDs for a type II LLE described by UNIQUAC and NRTL. Details can be found in Appendix F. The type I LLE is the one based on experimental data of the ternary system water(1)-acetone(2)-toluene(3), the type II LLE is based on experimental data of the ternary system 3-methyltetrahydrofurane(1) - di-n-butylether(2)

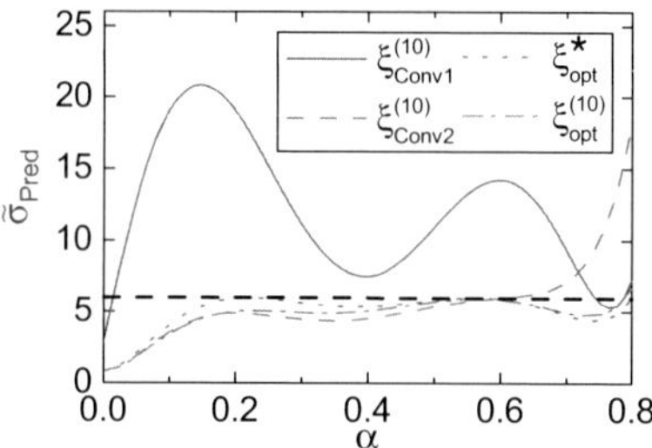

Figure 7.2: Predicted standardized variances for the optimal and conventional experimental designs plotted against the design variable α. The designs are based on UNIQUAC.

- water(3).

Figure 7.3(left) shows the positions of the optimal experiments for the type I LLE described by NRTL. Just as in the design for type I LLE described by UNIQUAC (cf. Section 7.2.1), the optimal experiments concentrate on regions where the curvature of the binodal curve is pronounced.

Figure 7.3 also shows the positions of the optimal experiments for the type II LLE described by NRTL (center) and UNIQUAC (right). Again, the optimal experiments concentrate on regions where the curvature of the binodal curve is pronounced.

To evaulate the overall quality of the optimal and conventional designs, we calculated the G- and D-efficiencies ζ_{G} and ζ_{D} (cf. Equation 3.23) as well as the mean predicted standardized variances mean($\tilde{\sigma}_{\mathrm{Pred}}$) (cf. Table 7.1).

The comparison of the continuous optimal designs $\boldsymbol{d}^{*}_{\mathrm{opt}}$ and the exact optimal designs $\boldsymbol{d}^{(10)}_{\mathrm{opt}}$ shows little loss of information in all cases. The predicted standardized variances of the optimal designs are low even in high curvature regions of the binodal curve. Again, this is due to the fact that the experiments concentrate on high curvature regions.

Regarding the G-efficiencies, we observe that the first conventional design $\boldsymbol{d}^{(10)}_{\mathrm{Conv1}}$ performs poorly in all cases: $\zeta_{\mathrm{G}}(\boldsymbol{d}^{(10)}_{\mathrm{Conv1}})$ does not exceed 50% for the designs based on NRTL, and is even less than 28% for the designs based on UNIQUAC, respectively.

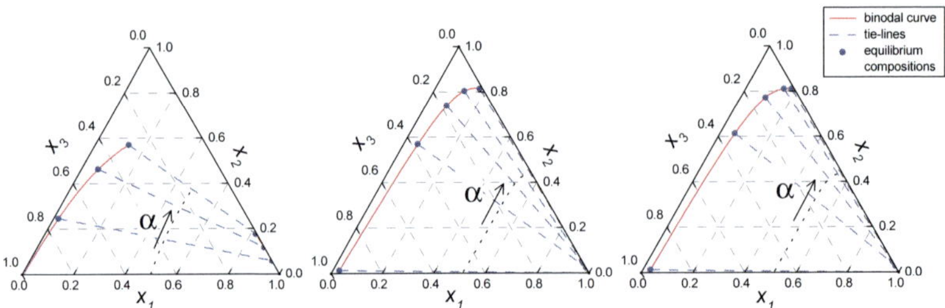

Figure 7.3: Positions of the optimally designed experiments in the ternary phase diagrams.
Left: $\alpha(\boldsymbol{d}_{\text{opt}}^{(10)})$ based on NRTL for type I LLE; center: $\alpha(\boldsymbol{d}_{\text{opt}}^{(10)})$ based on NRTL for type II LLE; right: $\alpha(\boldsymbol{d}_{\text{opt}}^{(10)})$ based on UNIQUAC for type II LLE.

For the conventional design based on UNIQUAC for a type II LLE, the G-efficiency $\zeta_{\text{G}}(\boldsymbol{d}_{\text{Conv1}}^{(10)})$ is particularly low (6%) due to an increased curvature of the binodal curve described by UNIQUAC compared to the binodal curve described by NRTL. The second conventional design $\boldsymbol{d}_{\text{Conv2}}^{(10)}$ also shows low values of G-efficiencies: $\zeta_{\text{G}}(\boldsymbol{d}_{\text{Conv2}}^{(10)}) \leq 37\%$. The G-efficiencies are especially low for the conventional designs applied to type II LLE described by UNIQUAC. This is due to very high predicted standardized variances in regions where the curvature of the binodal curve is large. In these cases, the prediction accuracy is increased by a factor up to 14 by the use of the optimal design. Still, the second conventional design $\boldsymbol{d}_{\text{Conv2}}^{(10)}$ performs well in regions where the curvature of the binodal curve is low.

The values of the D-efficiencies of the first conventional design, $\zeta_D\left(\boldsymbol{d}_{\text{Conv1}}^{(10)}\right)$, are between 55% and 93%, the values of the D-efficiencies of the second conventional design, $\zeta_D\left(\boldsymbol{d}_{\text{Conv2}}^{(10)}\right)$, are between 63% and 78%. Hence, with regard to the parameters, less information is lost than for the predictions when conventional designs are used instead of optimal designs. The parameter accuracy can thus be increased by a factor up to two by the use of an optimal experimental design.

According to the general equivalence theorem (cf. equation 3.22), the maximum value of the predicted standardized variance does not exceed the value of the number of parameters (i.e. six for a g^{E} model describing a ternary system) for a continuous optimal experimental design. As this is the lowest maximum value that can be achieved, it is a suitable benchmark to compare the mean values of the predicted standardized variances of the different designs to each other. The first conventional design $\boldsymbol{d}^{(10)}_{\text{Conv1}}$ shows the largest mean predicted standardized variance in all cases. These largest mean predicted standardized variance result from the distinct maxima of the mean predicted standardized variances, which are located in between the experimental supporting points (cf. figure 7.2). The second conventional design $\boldsymbol{d}^{(10)}_{\text{Conv2}}$ shows mean values of the predicted standardized variances that are in the range of and even slightly lower than the values of the optimal designs. This is due to very low values of the predicted standardized variances where the curvature of the binodal curve is low as a consequence of the many distinct experiments of $\boldsymbol{d}^{(10)}_{\text{Conv2}}$.

Table 7.1: G- and D-efficiencies ζ_{G} and ζ_{D} (equation 3.23) and mean predicted standardized variances mean(d_{Pred}) (equation 3.19) of the conventional designs $\boldsymbol{d}^{(10)}_{\text{Conv1}}$, $\boldsymbol{d}^{(10)}_{\text{Conv2}}$ and of the optimal designs $\boldsymbol{d}^{*}_{\text{opt}}$, $\boldsymbol{d}^{(10)}_{\text{opt}}$ of type I and type II LLE based on NRTL and UNIQUAC.

	Type I							
g^{E} model	UNIQUAC				NRTL			
Design $\boldsymbol{d}$	$\boldsymbol{d}^{(10)}_{\text{Conv1}}$	$\boldsymbol{d}^{(10)}_{\text{Conv2}}$	$\boldsymbol{d}^{*}$	$\boldsymbol{d}^{(10)}_{\text{opt}}$	$\boldsymbol{d}^{(10)}_{\text{Conv1}}$	$\boldsymbol{d}^{(10)}_{\text{Conv2}}$	$\boldsymbol{d}^{*}$	$\boldsymbol{d}^{(10)}_{\text{opt}}$
ζ_{G}	0.26	0.33	1	0.89	0.43	0.33	1	0.72
ζ_{D}	0.62	0.67	1	0.99	0.75	0.70	1	0.99
mean$(\tilde{\sigma}_{\text{Pred}})$	12.07	5.47	5.03	4.75	7.53	5.45	5.22	5.69

	Type II							
g^{E} model	UNIQUAC				NRTL			
Design $\boldsymbol{d}$	$\boldsymbol{d}^{(10)}_{\text{Conv1}}$	$\boldsymbol{d}^{(10)}_{\text{Conv2}}$	$\boldsymbol{d}^{*}$	$\boldsymbol{d}^{(10)}_{\text{opt}}$	$\boldsymbol{d}^{(10)}_{\text{Conv1}}$	$\boldsymbol{d}^{(10)}_{\text{Conv2}}$	$\boldsymbol{d}^{*}$	$\boldsymbol{d}^{(10)}_{\text{opt}}$
ζ_{G}	0.06	0.24	1	0.83	0.36	0.34	1	0.88
ζ_{D}	0.55	0.63	1	0.99	0.79	0.74	1	0.99
mean$(\tilde{\sigma}_{\text{Pred}})$	14.08	5.75	5.57	6.09	6.42	5.20	5.39	5.57

7.3 Conclusion

In this chapter, the theoretical potential of model-based optimal experimental designs (OED) for the characterization of type I and type II LLE has been investigated. The designs are based on the g^{E} models NRTL and UNIQUAC, since they are widely used in liquid-liquid phase equilibrium calculations.

The results show that simply reducing the number of distinct experimental settings and distributing them equally leads to poor predictions in regions where no measurements have been performed, especially if those regions include an increased curvature of the binodal curve. Simply increasing the number of equidistanly distibuted supporting points in conventional designs leads to good predictions in most parts of the phase diagram. However, this approach results in high experimental effort and loss of performance in high-curvature regions of the binodal curve.

The investigations show that optimal experimental designs for the characterization of LLE employ very few (3-5) distinct experiments and that a large fraction of the experimental effort is located in regions where the curvature of the binodal curve is high. Thereby, at least half of the experimental effort can be saved without losing neither prediction nor parameter accuracy.

Chapter 8

Optimal experimental design of diffusion and liquid-liquid equilibria experiments for extraction processes

In Chapters 5 and 7, optimal designs for diffusion and LLE experiments have been determined. When using these optimal designs, significant amounts of experimental effort can be saved in comparison to conventional, non-optimal designs. The optimal experimental designs aim for the precise determination of model parameters, i.e. diffusion coefficients and g^{E} model parameters.

However, in industrial applications, the precise knowledge of model parameters is typically not of main interest. Instead, operational and investment costs and the uncertainty in costs are of main interest. Costs and uncertainty in costs strongly influence investment decisions. Therefore, both costs and uncertainty in costs should be as low as possible. The costs depend on the sizing of equipment which in turn depends on the physical properties, i.e. the model parameters. Thus, uncertainties in the model parameters affect uncertainties in costs.

However, different physical properties may have different impact on equipment sizing and costs. When costs and uncertainty in costs should be minimized, only those physical property experiments should be performed that have significant impact on costs. OED techniques can identify these relevant physical property experiments.

OED with regard to the application has been investigated in a number of works. Recker et al. (2013) determined optimal kinetic experiments for a reaction-distillation network. Recker et al. (2012) calculated OED for batch distillation processes and semi-batch reactors. Houska et al. (2015) and Telen et al. (2014, 2015) investigated optimal designs for dynamic experiments with bioreactors. Bernardo et al. (2000, 2001, 2003) and Bernardo and Saraiva (2004, 2015) investigated the trade-off between costs of experiments and the uncertainty of process costs. Walz et al. (2020) developed a tailored OED method which aims for mitigating the worst-case realization of process equipment. The results of these studies can elucidate the real impact of different physical properties on process applications.

Extraction is a process application that is determined by both LLE and diffusion: While LLE determine the extraction grade that is principally accessible, diffusion determines the rate, i.e. the speed of the extraction process. The question at hand is: Which experiments provide more benefit for a reliable dimensioning of extraction processes with high confidence in costs: diffusion or LLE experiments?

In this chapter, we investigate the influence of diffusion and LLE experiments on the accuracy of the cost estimation of extraction processes. The investigation is based on a rate-based extraction process model and a cost estimation model. We identify optimal experimental designs minimizing the uncertainty in cost estimation of extraction processes. By comparison of the optimal designs with the designs derived in Chapters 5 and 7, we identify common features and differences in optimal experimental designs for applications and for physical property measurements. Thereby, general design rules are identified.

8.1 Extraction process model

Both equilibrium and mass transport properties determine the efficiency of an extraction process. Typical short-cut models for extraction processes are based on equilibrium properties only (Gmehling et al., 2012). These short-cut models are well-suited as long as mass transport does not introduce any limitation into the extraction process. In general, a rate-based mass transfer model is needed that incorporates both equilibrium and mass transport properties. This rate-based model is the basis for column sizing used in this work which determines the total annual costs of the extraction process.

8.1.1 Rate-based mass transfer model

We consider a counter-current extraction process with a continuous raffinate phase " and a dispersed extract phase '. We assume that extract and raffinate are non-soluble within each other. This assumption is well justified for systems with large miscibility gap. Therefore, calculations are based on molar loadings X_2 of mol transition component per mol extract carrier and raffinate carrier, respectively:

$$X_2'' = \frac{x_2''}{1 - x_2''} \tag{8.1}$$

$$X_2' = \frac{x_2'}{1 - x_2'}. \tag{8.2}$$

The total mole balance of the extraction process reads as (cf. Figure 8.1(a))

$$\dot{n}_1''(X_{2,\text{out}}'' - X_{2,\text{in}}'') = \dot{n}_3'(X_{2,\text{in}}' - X_{2,\text{out}}'). \tag{8.3}$$

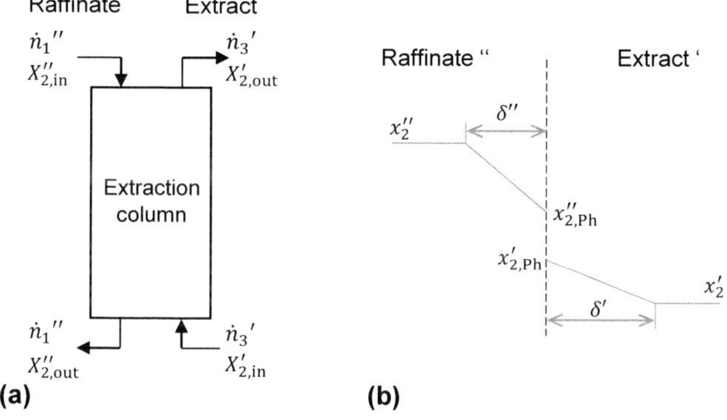

Figure 8.1: (a) Schematic mass balance of the countercurrent extraction profile. (b) Mole fraction profile of the transferred component 2 at the interface of raffinate and extract based on the two film theory.

Here, $\dot{n}_1''$ is the molar flow of the raffinate carrier component, and $X_{2,\text{out}}''$ and $X_{2,\text{in}}''$ are the outlet and inlet loadings of the raffinate. $\dot{n}_3'$ is the molar flow of the extract carrier component, and $X_{2,\text{in}}'$ and $X_{2,\text{out}}'$ are the inlet and outlet loadings of the extract.

The differential mole balance for the raffinate phase ' reads as:

$$\dot{n}_1' \mathrm{d}X_2'' = -\dot{N}_2 \mathrm{d}A \tag{8.4}$$

Thus, the differential depletion $\dot{n}_1' \mathrm{d}X_2''$ of the raffinate phase from the transferred component is equal to the molar flux $\dot{N}_2$ of the transferred component towards the extract phase through the differential mass transfer area $\mathrm{d}A$. $\mathrm{d}A$ can be described in terms of the specific mass transfer area a_P of the column internals, the column diameter d_K, and the differential height $\mathrm{d}z$:

$$\mathrm{d}A = a_\mathrm{P} \frac{\pi d_K^2}{4} \mathrm{d}z. \tag{8.5}$$

The molar flux $\dot{N}_2$ is proportional to the driving force, which is the difference in mole fraction of transferred component in the bulk phase, x_2'', and at the interface between raffinate and extract phase, $x_{2,\text{Ph}}''$ (cf. Figure 8.1(b)). Due to the fact that only the transition component is transferred through the interface, the interface acts like a semipermeable membrane. Therefore, a Stefan-correction has to be introduced in the formulation for the driving force (for details, the reader is referred to Taylor and Krishna (1993)). The transport equation then reads as

$$\dot{N}_2 = \frac{\beta_{12}}{V''} \ln \frac{1 - x_{2,\text{Ph}}''}{1 - x_2''} = \frac{\beta_{12}}{V''} \ln \frac{1 + X_2''}{1 + X_{2,\text{Ph}}''}. \tag{8.6}$$

Here, V'' is the molar volume of the raffinate phase, and β_{12} is the mass transfer coefficient. β_{12} is assumed to be independent of composition. β_{12} follows from the diffusion coefficient D_{12} and the thickness of the boundary layer of the raffinate phase, δ'':

$$\beta_{12} = \frac{D_{12}}{\delta''}. \tag{8.7}$$

Typically, the thickness of the boundary layer is often unknown (Taylor and Krishna, 1993). Therefore, the mass transfer coefficient β_{12} is typically calculated directly from empirical correlations employing the Sherwood number (Taylor and Krishna, 1993).

For the qualitative analysis pursued in this chapter, we want to analyze the impact of the diffusion coefficient D_{12}. Therefore, we assume the thickness of the boundary layer to be known and constant. If required, the thickness of the boundary layer could be re-calculated from empirical Sherwood number correlations and incorporated into the model.

Combining Equations 8.4, 8.5, 8.6 and integration leads to

$$\int_{X''_{2,\mathrm{in}}}^{X''_{2,\mathrm{out}}} \frac{\mathrm{d}X''_2}{\ln \frac{1+X''_2}{1+X''_{2,\mathrm{Ph}}}} = -\int_0^{h_\mathrm{K}} \frac{\beta_{12} a_\mathrm{P} \left(\frac{\pi d_\mathrm{K}^2}{4}\right)}{V'' \dot{n}''_1} \mathrm{d}z. \tag{8.8}$$

which can be reformulated into the well-known HTU-NTU-formulation

$$h_\mathrm{K} = \underbrace{\frac{V'' \dot{n}''_1}{\beta_{12} a_\mathrm{P} \left(\frac{\pi d_\mathrm{K}^2}{4}\right)}}_{HTU} \underbrace{\int_{X''_{2,\mathrm{out}}}^{X''_{2,\mathrm{in}}} \frac{\mathrm{d}X''_2}{\ln \frac{1+X''_2}{1+X''_{2,\mathrm{Ph}}}}}_{NTU}. \tag{8.9}$$

Here, HTU stands for "height of a transfer unit" and NTU stands for "number of transfer units".

To calculate the number of transfer units NTU, the loading $X''_{2,\mathrm{Ph}}$ at the interface is needed. We assume thermodynamic equilibrium at the interface, i.e. (c.f. Section 2.2.2)

$$T'_\mathrm{Ph} = T''_\mathrm{Ph} \tag{8.10}$$

$$p'_\mathrm{Ph} = p''_\mathrm{Ph} \tag{8.11}$$

$$x'_{2,\mathrm{Ph}}\, \gamma'_{2,\mathrm{Ph}} = x''_{2,\mathrm{Ph}}\, \gamma''_{2,\mathrm{Ph}}. \tag{8.12}$$

The molar flux $\dot{N}_2$ of the transferred component is absorbed by the extract phase. The driving force within the boundary layer of the extract phase is the difference in mole fractions of the transition component between at the interface of the extract phase, $x'_{2,\mathrm{Ph}}$, and in the bulk phase of the extract, x'_2 (cf. Figure 8.1(b)). Again, the Stefan-correction has to be considered. Thus, an expression for $\dot{N}_2$ in the extract phase is obtained analogously to Equation 8.6:

$$\dot{N}_2 = -\frac{\beta_{23}}{V'} \ln \frac{1 - x'_{2,\mathrm{Ph}}}{1 - x'_2} = \frac{\beta_{23}}{V'} \ln \frac{1 + X'_2}{1 + X'_{2,\mathrm{Ph}}}. \tag{8.13}$$

Here,

$$\beta_{23} = \frac{D_{23}}{\delta''} \tag{8.14}$$

is the (constant) mass transfer coefficient of acetone in the extract phase.

By dividing Equations 8.6 and 8.13, a relation between bulk phase loadings and loadings at the interface between raffinate and extract is obtained:

$$-\frac{V''\beta_{12}}{V'\beta_{23}} = \frac{\ln \frac{1+X'_2}{1+X'_{2,\mathrm{Ph}}}}{\ln \frac{1+X''_2}{1+X''_{2,\mathrm{Ph}}}}. \tag{8.15}$$

The model Equations 8.1, 8.2, 8.3, 8.7, 8.9, 8.10, 8.11, 8.12, 8.14, 8.15 enable the calculation of the concentration profiles inside the extraction column and the calulation of the column height h_K.

The column diameter d_K is calculated from the maximum volume flows of the raffinate and extract phase, respectively (Strigle, 1994):

$$\frac{\pi d_\mathrm{K}^2}{4} = \left(\underbrace{\frac{\dot{n}''_2 V''(1 + \max(X''))}{u''}}_{\text{max. volume flow of raffinate}} + \underbrace{\frac{\dot{n}'_3 V'(1 + \max(X'))}{u'}}_{\text{max. volume flow of extract}} \right) \cdot S. \tag{8.16}$$

Here, $S = 5$ is a safety factor that ensures sufficient distance of the operating point from flooding (Strigle, 1994).

The maximum volume flows of the raffinate and extract phase depend on the flooding velocities of the raffinate and extract phase, u' and u''. u' and u'' are calculated from the Crawford-Wilke flooding correlation (Strigle, 1994)

$$\frac{\rho'}{a_\mathrm{P}\mu''} \left(\sqrt{u'} + \sqrt{u''}\right)^2 = f\left(\frac{\mu'}{|\rho'' - \rho'|} \left[\frac{\sigma_\mathrm{I}}{\rho''}\right]^{0.2} \left[\frac{a_\mathrm{P}}{\epsilon}\right]^{1.5} \right). \tag{8.17}$$

Here, ρ'' and ρ' are the densities of the raffinate and extract phase, μ'' is the viscosity of the continuous phase (which is the raffinate phase in this case), σ_I is the interfacial tension, and ϵ is the void faction of the column packing. The function f is taken from a plot given by Strigle (1994).

In summary, the rate-based mass transfer model enables the calculation of the column dimensions as a function of the inlet conditions and the purification yield of the raffinate, the inlet conditions of the extract, and the physical properties of the system. The column dimensions determine the investment costs of the extraction column whereas the molar fluxes determine the annual operational costs of the extraction process. In the next section, the cost estimation model is introduced in detail.

8.1.2 Cost estimation model

The property of interest for the owner of the extraction column are the total annual costs TAC. The TAC result from total annual operational costs $TAOC$ and the depreciation of the investment costs, the total updated bare module costs $TUBMC$ (Biegler et al., 1997):

$$TAC = TAOC + ccf \cdot TUBMC. \tag{8.18}$$

ccf is the capital charge factor which determines the depreciation :

$$ccf = \frac{(q+1)^{N_{\text{life}}} \cdot q}{(q+1)^{N_{\text{life}}} - 1}, \tag{8.19}$$

where N_{life} is the lifetime of the extraction column and q is the interest rate. The total annual operational costs $TAOC$ are approximated by the disposal costs for the extract:

$$TAOC = (\dot{n}_3'(1 + \max(X'))) \cdot t_{\text{a}} \cdot C_{\text{disp}} \tag{8.20}$$

Here, t_{a} is the annual operational time, and C_{disp} are the specific disposal costs. Thus, the total annual operational costs $TAOC$ depend on the extract flow rate calculated from the rate-based mass transfer model.

The total updated bare module costs $TUBMC$ are calculated following the systematic approach of Guthrie (Biegler et al., 1997):

$$TUBMC = UF \cdot (BC(MPF + MF - 1)) \tag{8.21}$$

Here, UF is an update factor which considers inflation. The base costs BC are determined by the column dimensions d_K and h_K that have been calculated from the rate-based mass transfer model:

$$BC = C_0 \left(\frac{d_\mathrm{K}}{d_\mathrm{K,0}}\right)^{\alpha_\mathrm{d}} \left(\frac{h_\mathrm{K}}{h_\mathrm{K,0}}\right)^{\alpha_\mathrm{h}} \tag{8.22}$$

Here, C_0 is a base investment, and $d_\mathrm{K,0}$ and $h_\mathrm{K,0}$ are reference dimensions for the diameter and height of the extraction column. α_d and α_h are empirical parameters.

MPF is a material and pressure factor which is determined by

$$MPF = F_\mathrm{m} \cdot F_\mathrm{p}, \tag{8.23}$$

where F_m is a material factor, and F_p is a pressure factor. Values for F_m and F_p are given in Appendix G.

The module factor MF is a correction of the base costs BC that considers large-scale savings when large investments BC are made:

$$MF = f(BC). \tag{8.24}$$

The functional relationship f is given in Appendix G.

8.2 Optimal experimental design calculation

The aim of OED for extraction processes is to determine which diffusion and LLE experiments should be performed to minimize the variance in the total annual costs TAC. The influcence of uncertain diffusion and LLE data on the total annual costs is two-fold: On the one hand, uncertain LLE data lead to uncertain solubilities and therefore to uncertainty in the required extract flow rate. Thereby, uncertain LLE data determine the uncertainty in the operational costs. On the other hand, uncertain diffusion and LLE data lead to uncertainty in the required column dimensions. Thereby, uncertain diffusion and LLE determine the uncertainty in the investment costs.

The uncertainty in diffusion and LLE data arises from measurment errors in diffusion and LLE experiments. In this chapter, we consider diffusion experiments in closed cells (or microfluidic channels) with measurements at one position and one time as introduced in Chapter 5 (any other diffusion geometry and measurement technique could be considered as well). For LLE experiments, we consider the LLE experiments introduced in Section 2.2. In both diffusion and LLE experiments, compositions x_i are measured to determine the model parameters $\boldsymbol{\theta}$, i.e. g^{E} model parameters and diffusion coefficients, respectively. The measured mole fractions are burdened with uncertainty $\boldsymbol{V}_{\boldsymbol{w}}$. We assume the measurements to be independent of each other with measurment uncertainty $\sigma_x = 1\,\text{mol-\%}$. Therefore, the measurement variance matrix $\boldsymbol{V}_{\boldsymbol{w}}$ is assumed to be diagonal with equal diagonal entries:

$$\boldsymbol{V}_w = \boldsymbol{I} \cdot \sigma_x^2. \tag{8.25}$$

The measurement uncertainties $\boldsymbol{V}_{\boldsymbol{w}}$ lead to uncertainties $\boldsymbol{V}_{\boldsymbol{\theta}}$ in the model parameters. For our design calculations, we assume the parameters to be known: $\boldsymbol{\theta} = \hat{\boldsymbol{\theta}}$. If the parameters are unknown, predictive methods or initial experiments could give an initial estimate of the parameters.

The parameter (co-)variance matrix $\boldsymbol{V}_{\hat{\boldsymbol{\theta}}}$ can be calculated from the Fisher information matrix (cf. Equation 3.7). Both diffusion and LLE experiments are independent of each other. Therefore, the Fisher information matrix is a diagonal block matrix:

$$\boldsymbol{F} = \begin{pmatrix} \boldsymbol{F}_{\text{LLE}} & 0 & 0 \\ 0 & \boldsymbol{F}_{\text{Diffusion'}} & 0 \\ 0 & 0 & \boldsymbol{F}_{\text{Diffusion''}} \end{pmatrix}. \tag{8.26}$$

The LLE Fisher information matrix $\boldsymbol{F}_{\text{LLE}}$ was introduced in Chapter 7; the diffusion Fisher information matrix for a closed diffusion cell $\boldsymbol{F}_{\text{Diffusion}}$ was introduced in Chapter 5.

The parameter uncertainty $\boldsymbol{V}_{\hat{\boldsymbol{\theta}}}$ in turn leads to uncertainty in the total annual costs, σ_{TAC}^2 (cf. Equation 3.15). To determine diffusion and LLE experiments minimizing σ_{TAC}^2, we have to determine a c-optimal design (cf. Equations 3.16 and 3.20):

$$\boldsymbol{d}_{c,\text{opt}} = \underset{\boldsymbol{d}}{\operatorname{argmax}} \left(\sigma_{TAC}^2(\boldsymbol{d}, \hat{\boldsymbol{\theta}})\right)^{-1}. \tag{8.27}$$

Noteworthy, the c-optimal design minimizes only the uncertainty σ^2_{TAC} in the total annual costs and not the total annual costs TAC itself. However, for a reasonable operation of the extraction column, the total annual costs TAC should be low. Therefore, we perform a 2-step optimization procedure: In the first step, we minimize the total annual costs TAC. The total annual costs TAC are determined by the physical properties of the extraction system, i.e. the parameters $\hat{\boldsymbol{\theta}}$, and the extract flow rate $\dot{n}'_3$. With regard to the extract flow rates,

- low extract flow rates $\dot{n}'_3$ lead to low disposal costs, but a high and thin column with high investment costs;
- large extract flow rates $\dot{n}'_3$ lead to high disposal costs and a low, but thick column with high investment costs.

Thus, there is an optimum extract flow rate $\dot{n}'_3$ leading to minimum TAC:

$$TAC_{\text{min}} = \underset{\dot{n}'_3}{\text{argmin}}\ TAC(\hat{\boldsymbol{\theta}}, \dot{n}'_3). \tag{8.28}$$

In the second optimization step, we determine the c-optimal design $\boldsymbol{d}_{\text{c,opt}}$ for the operation point with minimum TAC_{min}. Of course, the c-optimal design could be determined for any other operation point as well.

8.3 Results

We consider a counter-current extraction process with the system water(1)-acetone(2)-toluene(3): acetone(2) is to be stripped of the raffinate consisting of water(1) and acetone(2) by means of the extracting agent toluene(3). The raffinate inlet flow is $5\,\text{kg/s}$ with a loading of acetone $X''_{2,\text{in}} = 0.2\,\text{mol acetone/mol water}$. The purification yield is $X''_{2,\text{out}} = 0.05\,\text{mol acetone/mol water}$. All process and cost parameters are given in Appendix G. The LLE data are modeled with UNIQUAC (results with NRTL are given in Appendix G).

The continuous c-optimal design minimizing the variance in the total annual costs

TAC is

$$
\boldsymbol{d}^{*}_{\text{c,opt}} = \left\{ \begin{array}{ccc|c|c} & \text{LLE} & & \text{Diffusion}' & \text{Diffusion}'' \\ \hline \alpha = 0.16 & \alpha = 0.61 & \alpha = 0.8 & \begin{array}{c} Fo = 0.1 \\ \xi = 0 \end{array} & \begin{array}{c} Fo = 0.1 \\ \xi = 0 \end{array} \\ \hline v = 0.76 & v = 0.07 & v = 0.06 & v = 0.08 & v = 0.03 \end{array} \right\}. \tag{8.29}
$$

Thus, 3 distinct LLE experiments and 1 diffusion experiment for each the extract phase ' and the raffinate phase " should be performed to minimize the variance σ^2_{TAC} of the total annual costs. 76% of the experimental effort should be spent on LLE experiments at overall composition $\alpha = 0.16$, 7% at $\alpha = 0.61$, and 6% at $\alpha = 0.61$. In addition, 8% of the experimental effort should be spent on diffusion experiments with the extract phase ' at dimensionless measurement time $Fo = 0.1$ and at the measurement position $\xi = 0$, i.e. at the wall of the closed diffusion cell. 3% of the experimental effort should be spent on diffusion experiments with the raffinate phase " at $Fo = 0.1$ and $\xi = 0$.

The exact c-optimal design $\boldsymbol{d}^{(12)}_{\text{c,opt}}$ for 12 measurements is built by rounding the design weights:

$$
\boldsymbol{d}^{(12)}_{\text{c,opt}} = \left\{ \begin{array}{ccc|c|c} & \text{LLE} & & \text{Diffusion}' & \text{Diffusion}'' \\ \hline \alpha = 0.16 & \alpha = 0.61 & \alpha = 0.06 & \begin{array}{c} Fo = 0.1 \\ \xi = 0 \end{array} & \begin{array}{c} Fo = 0.1 \\ \xi = 0 \end{array} \\ \hline v = 8/12 & v = 1/12 & v = 1/12 & v = 1/12 & v = 1/12 \end{array} \right\}. \tag{8.30}
$$

Figure 8.2 shows the positions of the experiments belonging to the c-optimal design $\boldsymbol{d}^{*}_{\text{c,opt}}$ (respectively $\boldsymbol{d}^{(12)}_{\text{c,opt}}$) in a ternary phase diagram. The positions of the optimal LLE experiments are similar to the D-optimal designs identified for the exclusive determination of LLE parameters (cf. Figure 8.2 and Figure 7.1 (right)). Effort is spent on experiments in the operational range of the extraction column and at high curvature regions of the binodal curve. However, the design weights differ significantly: For the c-optimal design, most experimental effort is spent on LLE experiments with compositions close to the operation point of the extraction column. In contrast, the D-optimal LLE experiments for the exclusive determination of LLE-parameters concentrate on high curvature regions of the binodal curve towards the critical point

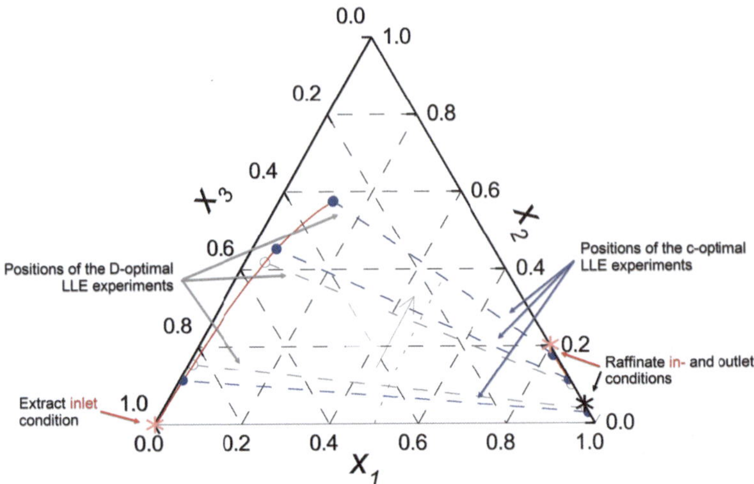

Figure 8.2: Positions of the c-optimal LLE experiments for the rate-based extraction process in the ternary phase diagrams. The LLE is based on UNIQUAC. For comparison, the positions of the D-optimal design are shown as well. In addition, the in- and outlet-conditions of the raffinate phase and the inlet condition of the extract phase is shown.

(cf. Equation 7.2).

To evaluate the impact of different designs on the uncertainty σ^2_{TAC} in the total annual costs, we calculate the c-efficiencies ζ_c (Equation 3.23). We consider designs with 12 measurements consisting of 10 LLE experiments (as in Chapter 7) and 1 diffusion experiment for each extract and raffinate phase. The diffusion experiments are always experiments at $Fo = 0.1$ and $\xi = 0$. For the positions and weights of the LLE experiments, different possibilities are investigated: We consider the c-optimal designs introduced above as well as the D-optimal and conventional designs for the exclusive determination of LLE parameters introduced in Chapter 7.

Table 8.1 shows the c-efficiencies of the considered designs:

- For the continuous c-optimal design $\boldsymbol{d}^*_{\mathrm{c,opt}}$ (Equation 8.29), the c-efficiency is

$\zeta_c = 100\%$ by definition.

- For the exact c-optimal design $\boldsymbol{d}^{12}_{\text{c,opt}}$ (Equation 8.30), the c-efficiency is $\zeta_c = 97\%$, i.e. little information is lost.
- If the exact D-optimal design $\boldsymbol{d}^{(12)}_{\text{opt,LLE}}$ for the exclusive determination of LLE parameters is applied (Equation 7.3), the c-efficiency is $\zeta_c = 66\%$, i.e. 34% of the information is lost and $1/0.66 = 1.5$ times more experiments would have to be performed to obtain the same accuracy in the total annual costs TAC as in the optimal design case.
- If the first conventional design $\boldsymbol{d}^{(12)}_{\text{Conv1}}$ is applied (Equation 7.4), almost all information is lost: the c-efficiency is only $\zeta_c = 4\%$, i.e. $1/0.04 = 25$ times more experiments would have to be performed to obtain the same accuracy in the TAC as in the optimal design case.
- If the second conventional design $\boldsymbol{d}^{(12)}_{\text{Conv2}}$ with 10 equidistantly distributed LLE experiments is applied (Equation 7.5), a moderate c-efficiency of $\zeta_c = 63\%$ is obtained. Thus, the second conventional design $\boldsymbol{d}^{(12)}_{\text{Conv2}}$ performs approximately as good as the exact D-optimal design $\boldsymbol{d}^{(12)}_{\text{opt,LLE}}$ for the exclusive determination of LLE parameters. However, whereas $\boldsymbol{d}^{(12)}_{\text{opt,LLE}}$ requires only 3 distinct LLE experiments, $\boldsymbol{d}^{(12)}_{\text{Conv2}}$ requires 10 distinct LLE experiments and is thus associated with increased experimental effort.

The comparison of the different designs reveals the large impact of experimental design on the accuracy of the total annual costs TAC. It is important to consider the application when experiments are performed: The optimal designs differ significantly when experiments are performed for accurate physical properties, e.g., to provide data for databanks, or when experiments are performed for process applications. The optimal designs should preferentially be calculated for every specific application. If required, robust design techniques should be applied (Pronzato and Walter, 1988). Simply reducing the number of distinct experiments without optimization can lead to almost useless designs (as shown by the performance of $\boldsymbol{d}^{(12)}_{\text{Conv1}}$).

Two characteristics of the c-optimal designs with UNIQUAC (Equation 8.29) and NRTL (cf. Appendix G) can be identified:

- LLE experiments are far more important than diffusion experiments to reduce the uncertainty σ^2_{TAC} in the total annual costs.

- Most LLE measurements are spent in a region of the miscibility gap where the actual extraction process takes place.

The characteristics seem to be reasonable since the properties of the LLE in the operation range of the extraction process are of major importance for the performance of the extraction process. Therefore, accurate knowledge of the LLE in the neighborhood of the operation point is important to reduce the uncertainty in the predicted total annual costs TAC.

Table 8.1: c-efficiencies ζ_c (Equation 3.23) of the continuous and exact c-optimal designs $\boldsymbol{d}^*$ and $\boldsymbol{d}_{\text{opt}}^{(12)}$, as well as of the D-optimal $\boldsymbol{d}_{\text{opt,LLE}}^{(12)}$ design for the exclusive determination of LLE parameters, and the conventional designs $\boldsymbol{d}_{\text{Conv1}}^{(12)}$ and $\boldsymbol{d}_{\text{Conv2}}^{(12)}$. The LLE is based on UNIQUAC

	$\boldsymbol{d}^*$	$\boldsymbol{d}_{\text{opt}}^{(12)}$	$\boldsymbol{d}_{\text{opt,LLE}}^{(12)}$	$\boldsymbol{d}_{\text{Conv1}}^{(12)}$	$\boldsymbol{d}_{\text{Conv2}}^{(12)}$
LLE design	(Eq. 8.29)	(Eq. 8.30)	(Eq. 7.3)	(Eq. 7.4)	(Eq. 7.5)
ζ_c	1	0.97	0.66	0.04	0.63

8.4 Conclusions

In this chapter, optimal experimental designs of diffusion and LLE experiments for the application in extraction processes have been identified. The experimental designs have been determined such that the variance in the total annual costs for the operation of an extraction column is minimized.

The identified optimal experimental designs concentrate on 3 distinct LLE experiments and a single diffusion experiment for each the raffinate and the extract phase. The optimal diffusion experiments coincide with results presented in Chapter 5. However, only a small fraction of experimental effort is spent on diffusion experiments. Most of the experimental effort is spent on LLE experiments with compositions which are in the operation range of the extraction process. However, effort is also spent on experiments at high curvature regions of the binodal curve which are outside of the operational range of the extraction column. This characteristic is in common with the optimal experimental LLE design employed for LLE parameter estimation (cf. Chapter 7).

Conventional designs with many equidistantly distributed supporting points perform at most as good as the designs optimized for physical parameter estimation. However, a multitude of experimental effort is required for these conventional designs.

Thus, optimal designs can save a significant amount of experimental effort while the accuracy of the predictions is even increased. To exploit the full potential of OED, experimental designs should be optimized for each individual application. Until then, a simple design rule can be derived: Experiments should be performed in the operational range of the considered application and at the optimal design points identified for physical parameter estimation.

CHAPTER 9

Summary and outlook

Diffusion and liquid-liquid equilibrium (LLE) experiments are burdened with long measurement times and large sample consumptions. In particular, multicomponent measurements require considerable experimental effort. These circumstances have led to the situation that there is a lack of experimental diffusion and LLE data. In addition, predictive models are not yet sufficiently accurate: Predictive LLE models are burdened with large uncertainty due to the to the complex molecular interactions; predictive diffusion models suffer from the lack of experimental data for validation purposes.

This thesis addresses the need for the fast and efficient generation of diffusion and LLE data. For this purpose, a model-based approach is applied to contribute to the development of predictive diffusion models and efficient diffusion and LLE experiments.

In Chapter 4, we performed Molecular Dynamics (MD) simulations of diffusion in non-ideal liquid binary Lennard-Jones systems. MD simulations offer two major advantages for the analysis of diffusion processes: (1) Due to recent progress in computational power, extensive datasets on diffusion can be calculated in a reasonable amount of time, and (2) MD simulations provide detailed insight into transport and equilibrium properties that is difficultly (and at most partially) obtainable from experiments. Based on the extensive MD dataset, we studied the composition-dependence of mutual and self-diffusion coefficients in non-ideal liquids. For mutual diffusion coefficients, the existing correlation of Moggridge that is based on the ideal Darken equation provides satisfactory predictions. However, the correlation of Moggridge depends on the knowledge of concentration-dependent self-diffusion coefficients, which are rarely available. Existing correlations do not provide satisfactory predictions of concentration-dependent self-diffusion coefficients. To close the gap, we developed a predictive equation for concentration-dependent self-diffusion coefficients in non-ideal binary liquids. The new predictive equation is validated with experimental data for binary mixtures with molar mass ratios $M_2/M_1 < 2$ and without dimerising species. The relative accuracy was found to increase from 10 % to 5 %. Thus, our new model provides the missing link to render Darken-based models into practical tools to predict mutual diffusion coefficients.

Even with increased accuracy of predictions, diffusion experiments are indispensable. To reduce the experimental effort of diffusion experiments, a reduction of the *number* of required diffusion experiments would be desirable if the accuracy of diffusion coefficients could be retained. This goal can be achieved with optimal experimental designs. In Chapter 5, we applied model-based experimental design techniques to identify optimal geometries that provide maximum information on diffusion coefficients. Both already existing and theoretically conceivable geometries are considered. Open geometries, where components can diffuse out of the geometry, are identified to be most beneficial. The most commonly employed free diffusion experiments are shown to lead to high uncertainty. Replacing free diffusion experiments by experiments with open geometries can improve the accuracy of diffusion coefficients by up to two orders of magnitude. Vice versa, the number of experiments could be reduced by two orders of magnitude without losing accuracy if optimal diffusion geometries are implemented.

Even the most sophisticated experimental design cannot overcome any limitation introduced by an experimental setup. Therefore, the model-based approach has to be combined with the development of the experimental setup. In Chapter 6, we apply the model-based approach to develop improved diffusion experiments for gases and liquids. In Section 6.1, the model-based experimental approach is applied to overcome long-standing discrepancies in the measurement of diffusion in gases with a Loschmidt cell. The model-based approach is shown to be a key tool for the identification of necessary improvements in both the model and the experimental setup. The new Loschmidt setup allows for the measurement of concentration-dependent binary diffusion coefficients in gases from a single experimental run. Thus, significant savings in experimental effort are achieved.

In Section 6.2, a microfluidic setup for the measurement of diffusion in liquids is developed that enables the measurement of multicomponent diffusion coefficients from a single experiment. Thus, a significant reduction of the number of experiments and the consumption of sample is achieved.

In Chapter 7, we identify optimal experimental designs for LLE measurements. The results show that optimal designs concentrate on only few (3 5) repeated LLE measurements at high-curvature regions of the binodal curve. These optimal designs yield accuracies of LLE predictions that are even better than using laborious equidistantly distributed LLE experiments. Thus, the application of optimal LLE designs enables a significant reduction of the number of LLE experiments without losing accuracy of the LLE model parameters and predictions.

Physical property measurements such as diffusion and LLE measurements are not an end in itself. They are used in process applications where not the accuracy of physical properties but the accuracy of other objectives such as, e.g. annual costs, is of major interest.
Liquid-liquid extraction is a process that depends on both diffusion and LLE. In Chapter 8, we analyzed the impact of experimental designs on the accuracy of cost estimations. Optimal designs for extraction concentrate on experiments similar to the designs for physical property measurements introduced in Chapters 5 and 7. However, only a small fraction of experimental effort should be spent on diffusion experiments. Most of the experimental effort should be spent on LLE experiments with compositions which are in the operation range of the extraction process. In addition, effort should be spent on experiments at high curvature regions of the binodal curve which are outside of the operational range of the extraction column. Conventional designs with many equidistantly distributed supporting points perform worse and require a multiple of experimental effort. Thus, optimal designs can save a significant amount of experimental effort while the accuracy of the predictions is even increased.

In summary, the applications presented in this thesis show that a model-based approach can contribute significantly to the development of efficient physical property measurements. The demonstration of the potential as well as the practical applications of the model-based approach provide a promising basis for future developments.

9.1 Perspective for future work

The results presented in this thesis give the opportunity for a number of future work:

Development of predictive engineering equations for engineering models

Our new engineering equation for the prediction of concentration-dependent self diffusion coefficients in non-ideal binary liquids provides a missing link for Darken-based models. However, the development of an extension for mixtures with dimerising species and molar mass ratios $M_2/M_1 > 2$ would be desirable. In addition, the study of multicomponent systems should be at the center of future investigations.

Development of optimal diffusion geometries

The optimal diffusion geometries identified in Chapter 5 include geometries that are not developed yet for diffusion of fluids. Nevertheless, the investigation shows the "route of development" that would be desirable. An idea for a future geometry could be, e.g., the development of a microfluidic device that provides two diffusion domains in the form of the closed cylinder. More creative ideas are certainly needed to develop further diffusion geometries.

Improvement of the Loschmidt setup

Although satisfying results are obtained from the improved Loschmidt setup, a small offset of 1.5 % in the diffusion coefficient remained. The thermal (in-)stability of the optical setup is assumed to be the reason for the offset. Therefore, the thermal stability should be checked and improved if necessary. In addition, the combination of the Loschmidt cell with measurement techniques that allow for a simultaneous composition analysis of multicomponent gas mixtures would be an interesting field of future development. Thereby, multicomponent diffusion coeffcicients of gases could be determined from a single experiment.

Development of microfluidic devices

The successful determination of multicomponent diffusion coefficients with a microfluidic device offers the potential for new microfluidic combinations. An example would be the combination of diffusion and LLE experiments into a single experiment. Thien et al. (2017) has already performed successful measurements of LLE in microfluidic devices. A combination of diffusion and LLE experiments would not only save substantial experimental effort, but would also yield effective mass transport coefficients that are required for extraction processes.

Optimal designs with regard to the application

The results on experimental designs for extraction processes give an interesting insight into the qualitative impact of diffusion and LLE measurements on costs. More rigorous rate-based models could be implemented to gain increased quantitative insight.
In general, more studies should be performed on optimal designs for applications: These investigations can serve as a reminder that the increase of accuracy of diffusion and LLE is not an end by itself. The development of improved experimental setups should follow the needs of applications. Efforts should be spent on the improvement of only those experiments that are *too* expensive and *not* sufficiently accurate.

Appendix A

A new predictive model for diffusion coefficients of nonideal liquids: Additional data

A.1 Details on the MD simulations

Our analysis is based on MD simulations of LJ systems. In the following, we provide a short overview on the specifications of the simulations. For more details and numeric results, the reader is referred to Jamali et al. (2018).

We performed 250 distinct MD simulations of binary LJ systems. All parameters and properties of these simulations are reported in reduced units. The parameters of the first species serve as base units: diameter $\sigma_1 = \sigma = 1$, interaction energy $\epsilon_1 = \epsilon = 1$, and mass $m_1 = m = 1$. The parameters of the second species and the adjustable parameter k_{ij} of the Lorentz-Berthelot mixing rule are listed in Table A.1. To cover a broad range of nonidealities, the ratios of the parameters of the first and second species are varied over a large range. The reduced temperature T and pressure p are $T = 0.65$ and $p = 0.05$. For each specified LJ system, two different types of simulations were performed: simulations to determine transport properties and simulations to determine thermodynamic factors.

Transport properties were calculated from equilibrium MD simulations with 200 million time steps with a time step length of 0.001 in reduced units. The transport coefficients were calculated from time-correlation functions. The Einstein relations were used to sample the time-correlations, i.e. the displacements of particles were sampled over time (Frenkel and Smit, 2002; Allen and Tildesley, 2017) (cf. Equations 2.24-2.26).

The values of the transport coefficients depend on the box size of the MD simula-

Table A.1: Specifications of the studied LJ systems. LJ particle of type 1 has $\sigma_1 = \sigma = 1.0$, $\epsilon_1 = \epsilon = 1.0$, and mass $= m_1 = 1.0$ in reduced units (Allen and Tildesley, 2017). k_{ij} is an adjustable parameter to the Lorentz-Berthelot mixing rule $\epsilon_{ij} = \sqrt{\epsilon_1 \epsilon_2}\,(1 - k_{ij})$, controlling the nonideality of the systems.

Specification	Values
Total number of particles	500, 1000, 2000, 4000
Independent simulations	10, 10, 5, 5
x_1	0.1, 0.3, 0.5, 0.7, 0.9
ϵ_2/ϵ_1	1.0, 0.8, 0.6, 0.5
σ_2/σ_1	1.0, 1.2, 1.4, 1.6
m_2/m_1	$(\sigma_2/\sigma_1)^3$
k_{ij}	0.05, 0.0, -0.3, -0.6

tions. More precisely, the transport coefficients scale linearly with the inverse of the box size, $1/L$. To correct for these finite-size effects, each LJ system was simulated for four different system sizes (500, 1000, 2000, and 4000 particles). Subsequently, the transport coefficients were extrapolated linearly to an infinite box size, i.e. $1/L \rightarrow 0$, to obtain the transport coefficients in the thermodynamic limit.

For the calculation of thermodynamic factors, equilibrium MD simulations with large systems consisting of 25000 particles were performed. The thermodynamic factors were calculated from Kirkwood-Buff coefficients (cf. Equation 4.2). The Kirkwood-Buff coefficients were calculated from integrals of the radial distribution functions (RDFs). Both the RDFs and the Kirkwood-Buff integrals were corrected for finite-size effects using the method of Ganguly and van der Vegt (2013) and Milzetti et al. (2018) for the RDFs and the method of Krüger et al. (2013), Dawass et al. (2017) and Krüger and Vlugt (2018) for the Kirkwood-Buff integrals. Each simulation for the calculation of thermodynamic factors was performed for 10 million time steps with a time step length of 0.001 in reduced units. All simulations were repeated for at least five times to assess statistical uncertainties.

A.2 References for the experimental data

Table A.2: References for the experimental data

System	Diffusion coefficients D_{12}	$D_{1,self}$	$D_{2,self}$	Thermodynamic factor Redlich-Kister (RK)	NRTL	reported in Literature (Lit)
Acetone-Benzene	Anderson et al. (1958), Cullinan and Toor (1965)	Yoshinobu and Yasumichi (1972)	Yoshinobu and Yasumichi (1972)	Moggridge (2012b)	Zhu et al. (2015)	-
Acetone-Carbon Tetrachloride	Anderson et al. (1958), Cullinan and Toor (1965)	Hardt et al. (1959)	Hardt et al. (1959)	Moggridge (2012b)	-	-
Acetone-Chloroform	McCall and Douglass (1967), Tyn and Calus (1975), Anderson et al. (1958)	D'Agostino et al. (2013)	D'Agostino et al. (2013)	D'Agostino et al. (2013)	Gmehling et al. (1979)	-
Acetone-Water	Anderson et al. (1958), Grossmann and Winkelmann (2005), Rehfeldt and Stichlmair (2010), Tyn and Calus (1975), Zhou et al. (2013)	Mills and Hertz (1980)	Mills and Hertz (1980)	Moggridge (2012b)	Gmehling et al. (1979)	-
Acetonitrile-Water	Easteal et al. (1987)	Easteal et al. (1987)	Easteal et al. (1987)	Fitted from data of French (1987)	-	-
Cyclohexane-Benzene	Harned (1957)	Mills (1965)	Mills (1965)	Moggridge (2012b)	-	-
Diethylether-Chloroform	Sanni et al. (1971), Weingärtner (1990)	Weingärtner (1990)	Weingärtner (1990)	Moggridge (2012b)	-	-
Ethanol-Benzene	Anderson et al. (1958), Zhu et al. (2015)	Johnson and Babb (1956)	Johnson and Babb (1956)	-	Zhu et al. (2015)	Guevara-Carrion et al. (2016b)*
Ethanol-Carbon Tetrachloride	Hammond and Stokes (1955), Longsworth (1966), Bosse and Bart (2005)	Hardt et al. (1959)	Hardt et al. (1959)	-	-	Guevara-Carrion et al. (2016b)*
Heptane-Benzene	Harris et al. (1970)	Harris et al. (1970)	Harris et al. (1970)	Moggridge (2012b)	-	-
Hexane-Benzene	Harris et al. (1970)	Harris et al. (1970)	Harris et al. (1970)	Moggridge (2012b)	-	-
Hexane-Toluene	Ghai and Dullien (1974)	Ghai and Dullien (1974)	Ghai and Dullien (1974)	Moggridge (2012b)	-	-
Methanol-Benzene	Caldwell and Babb (1956)	Aoyagi and Albright (1972), Johnson and Babb (1956)	Aoyagi and Albright (1972), Johnson and Babb (1956)	-	-	Guevara-Carrion et al. (2016b)*
Methanol-Carbon Tetrachloride	Anderson et al. (1958), Prabhakar and Weingärtner (1983), Longsworth (1966)	Prabhakar and Weingärtner (1983)	Prabhakar and Weingärtner (1983)	-	-	Guevara-Carrion et al. (2016b)*
Methanol-Water	Chang et al. (2005), Derlacki et al. (1985), Bosse and Bart (2005)	Derlacki et al. (1985)	Derlacki et al. (1985)	Moggridge (2012b)	-	-
Nitrobenzene-Hexane	Haase and Siry (1968)	D'Agostino et al. (2011)	D'Agostino et al. (2011)	D'Agostino et al. (2011)	-	-
Water-N-methylpyridine	Ambrosone et al. (1995)	Ambrosone et al. (1995)	Ambrosone et al. (1995)	Moggridge (2012b)	Zhu et al. (2015)	-

*MD simulation results verified with experimental data

A.3 Self-diffusion coefficients of Lennard-Jones (LJ) systems

A.3.1 Relative deviations $\Delta D_{2,\mathrm{self,rel}}$ of the McCarty-Mason prediction as a function of the thermodynamic factor Γ for component 2

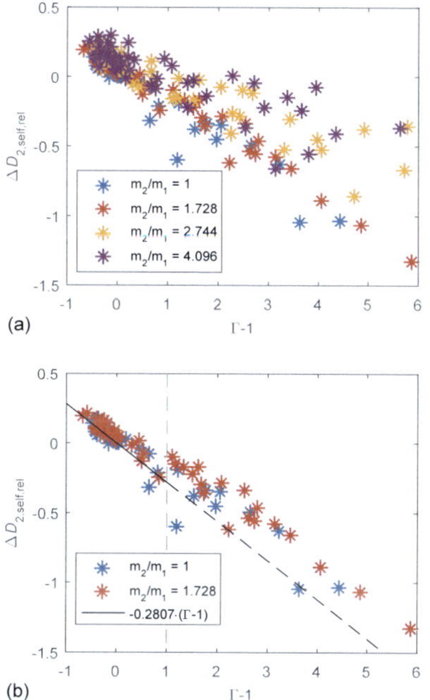

Figure A.1: Relative deviations $\Delta D_{2,\mathrm{self,rel}}$ of the McCarty-Mason prediction as function of the thermodynamic factor Γ for LJ systems.
(a) $\Delta D_{2,\mathrm{self,rel}}$ for all LJ systems, differentiated by the molar mass ratios m_2/m_1.
(b) $\Delta D_{2,\mathrm{self,rel}}$ for LJ systems with molar mass ratios $m_2/m_1 < 2$ and best fit of Equation 15 (black line) for $0 < \Gamma < 2$ (indicated by the vertical dashed line).

A.3.2 LJ systems with molar mass ratios $m_2/m_1 < 2$

Composition-dependent self-diffusion coefficients $D_{i,\mathrm{self}}$, thermodynamic factors $\Gamma - 1$, and relative deviations $\Delta D_{i,\mathrm{self,rel}}$ of LJ systems with molar mass ratios $m_2/m_1 < 2$. The specifications of the LJ systems ϵ_2/ϵ_1, σ_2/σ_1, m_2/m_1, and k_{ij} are given in the title of each figure.

Top figures: Blue stars: Simulation results of self-diffusion coefficients $D_{i,\mathrm{self}}$ of binary LJ systems as function of the mole fraction x_1 of the first species. Blue dashed line: smoothing fit to the simulation results; red circles/line: predictions of the McCarty-Mason equation (Equation 6); green diamonds/line: predictions of the modified McCarty-Mason equation (Equation 25). The error bars of $D_{i,\mathrm{self}}$ are smaller than the symbols in most cases. Please note that y-axes are adapted for each system.

Bottom figures: Composition dependence of the thermodynamic factor $\Gamma - 1$ (blue stars/line, left axis) and composition dependence of the relative deviation $\Delta D_{i,\mathrm{self,rel}}$ between the self-diffusion coefficients and the predictions of the McCarty-Mason equation (Equation 6) (red circles/line, right axis) and the modified McCarthy-Mason equation (Equation 25) (green diamonds/line, right axis). A clear correlation between $\Gamma - 1$ and $\Delta D_{i,\mathrm{self,rel}}$ can be observed. The error bars of $\Gamma - 1$ are smaller than the symbols in most cases. Please note that y-axes are adapted for each system.

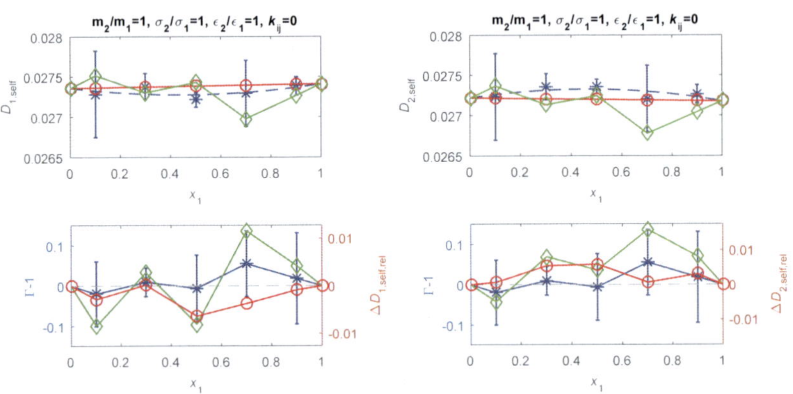

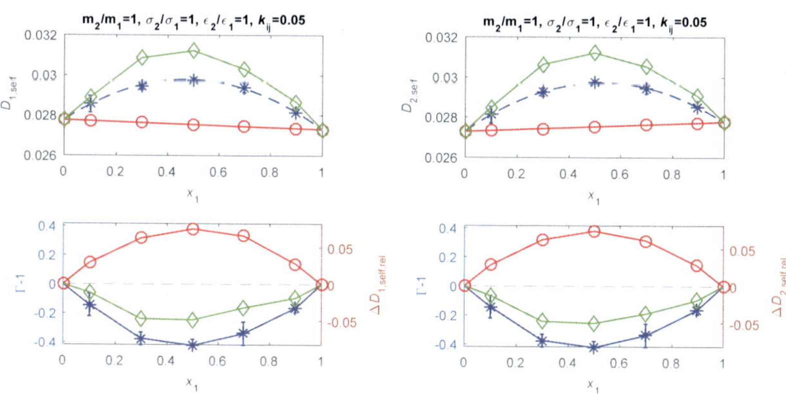
m2/m1=1, σ2/σ1=1, ε2/ε1=1, kij=0.05
D1,self
D2,self
x1
Γ-1
ΔD1,self,rel
ΔD2,self,rel

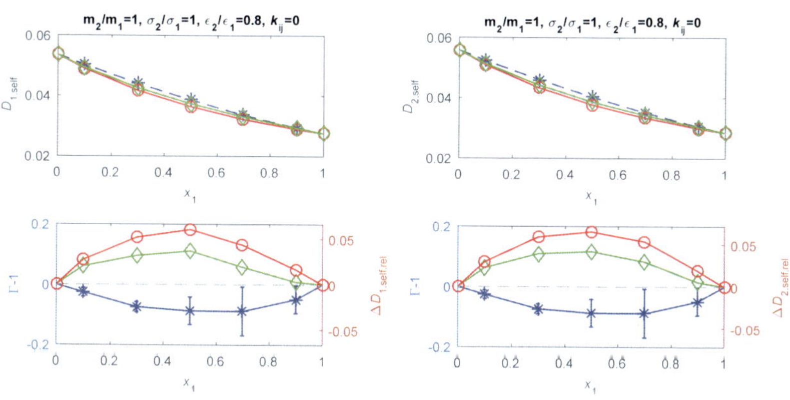
m2/m1=1, σ2/σ1=1, ε2/ε1=0.8, kij=0
D1,self
D2,self
x1
Γ-1
ΔD1,self,rel
ΔD2,self,rel

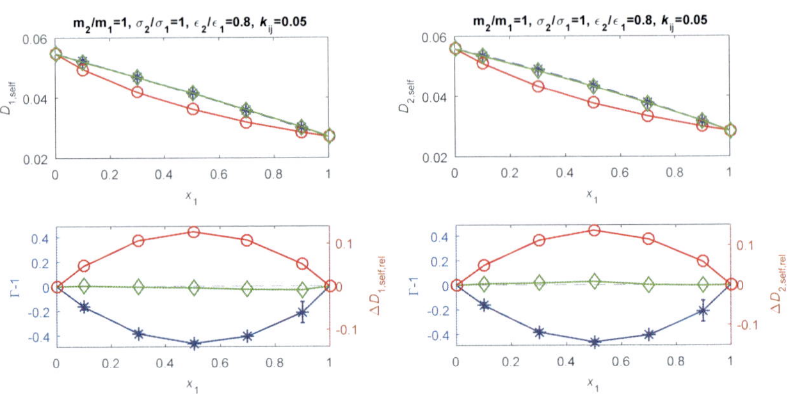
m2/m1=1, σ2/σ1=1, ε2/ε1=0.8, kij=0.05
m2/m1=1, σ2/σ1=1, ε2/ε1=0.8, kij=0.05
D1,self
D2,self
x1
Γ-1
ΔD1,self,rel
ΔD2,self,rel

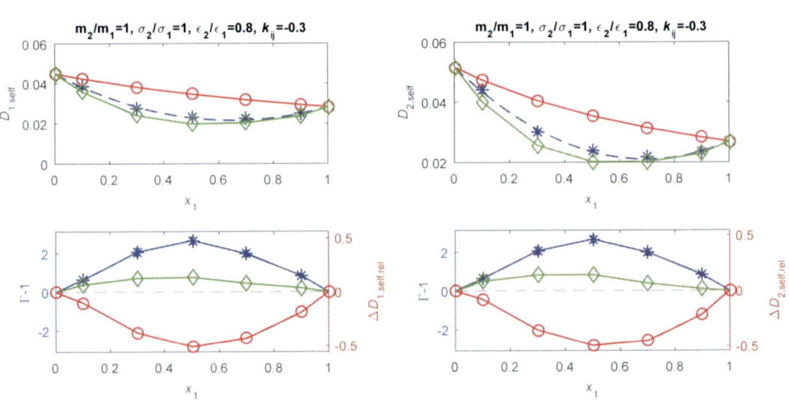
m2/m1=1, σ2/σ1=1, ε2/ε1=0.8, kij=-0.3
m2/m1=1, σ2/σ1=1, ε2/ε1=0.8, kij=-0.3
D1,self
D2,self
x1
Γ-1
ΔD1,self,rel
ΔD2,self,rel

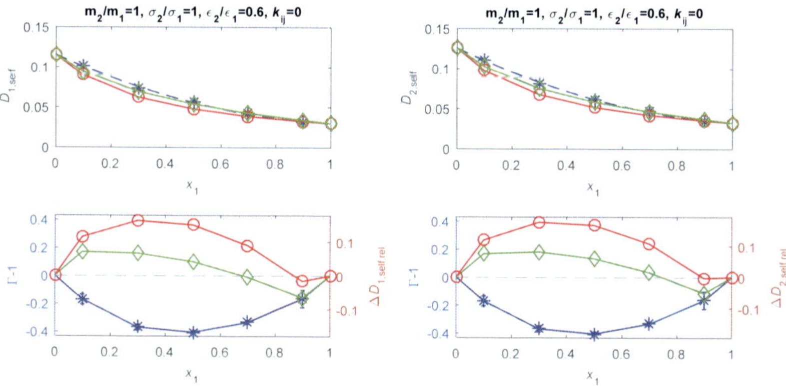
m_2/m_1=1, σ_2/σ_1=1, ϵ_2/ϵ_1=0.6, k_{ij}=0
$D_{1,self}$
x_1
Γ-1
$\Delta D_{1,self,rel}$
m_2/m_1=1, σ_2/σ_1=1, ϵ_2/ϵ_1=0.6, k_{ij}=0
$D_{2,self}$
x_1
Γ-1
$\Delta D_{2,self,rel}$

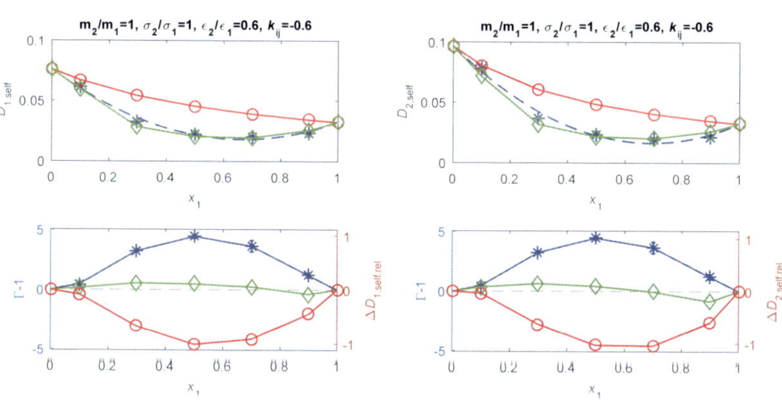
m_2/m_1=1, σ_2/σ_1=1, ϵ_2/ϵ_1=0.6, k_{ij}=-0.6
$D_{1,self}$
x_1
Γ-1
$\Delta D_{1,self,rel}$
m_2/m_1=1, σ_2/σ_1=1, ϵ_2/ϵ_1=0.6, k_{ij}=-0.6
$D_{2,self}$
x_1
Γ-1
$\Delta D_{2,self,rel}$

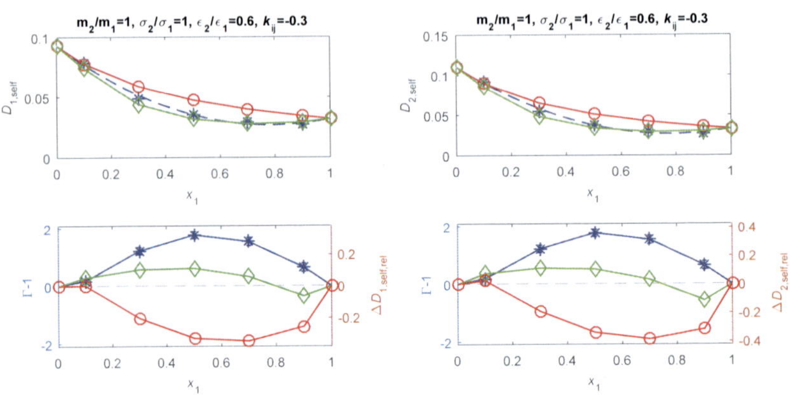

m2/m1=1, σ2/σ1=1, ε2/ε1=0.6, kij=-0.3
D1,self
x1
Γ-1
ΔD1,self,rel
m2/m1=1, σ2/σ1=1, ε2/ε1=0.6, kij=-0.3
D2,self
x1
Γ-1
ΔD2,self,rel

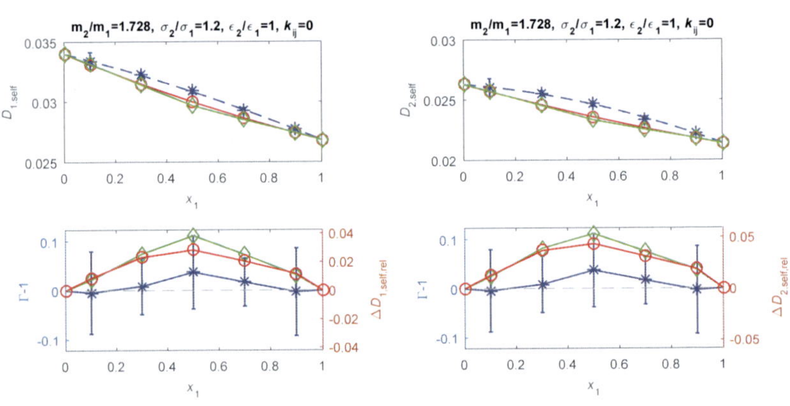

m2/m1=1.728, σ2/σ1=1.2, ε2/ε1=1, kij=0
D1,self
x1
Γ-1
ΔD1,self,rel
m2/m1=1.728, σ2/σ1=1.2, ε2/ε1=1, kij=0
D2,self
x1
Γ-1
ΔD2,self,rel

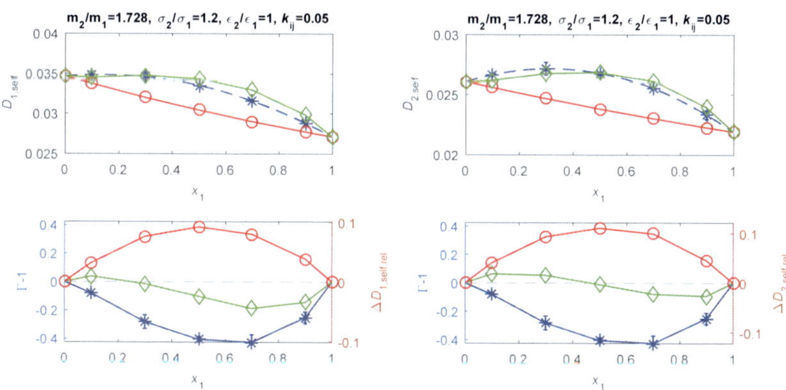

m2/m1=1.728, σ2/σ1=1.2, ε2/ε1=1, kij=0.05
D1,self
x1
Γ-1
ΔD1,self,rel
m2/m1=1.728, σ2/σ1=1.2, ε2/ε1=1, kij=0.05
D2,self
x1
Γ-1
ΔD2,self,rel

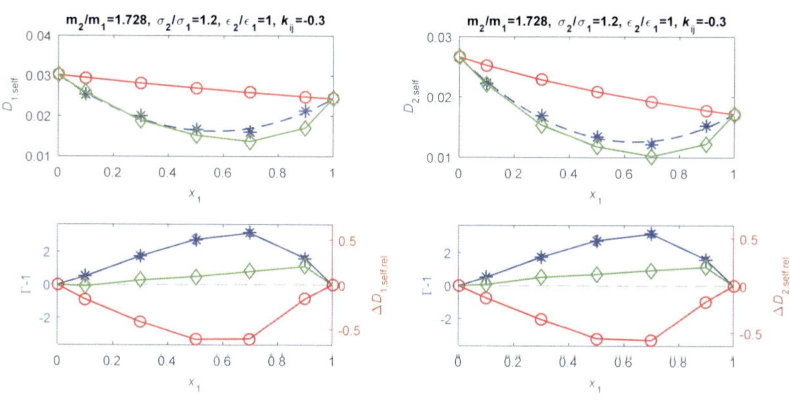

m2/m1=1.728, σ2/σ1=1.2, ε2/ε1=1, kij=-0.3
D1,self
x1
Γ-1
ΔD1,self,rel
m2/m1=1.728, σ2/σ1=1.2, ε2/ε1=1, kij=-0.3
D2,self
x1
Γ-1
ΔD2,self,rel

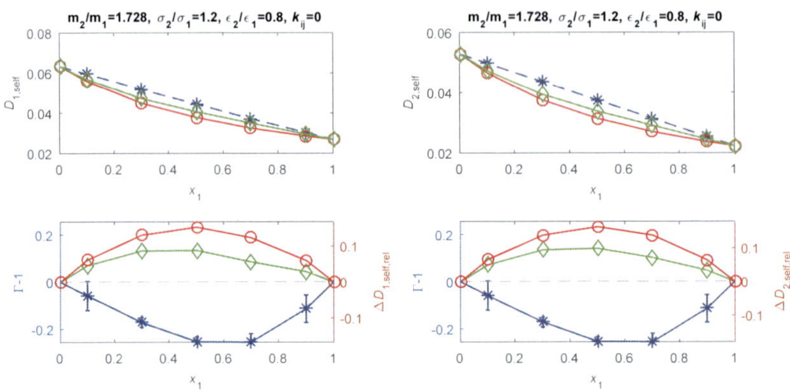
m_2/m_1=1.728, σ_2/σ_1=1.2, ϵ_2/ϵ_1=0.8, k_{ij}=0
$D_{1,self}$
$D_{2,self}$
x_1
Γ-1
$\Delta D_{1,self,rel}$
$\Delta D_{2,self,rel}$

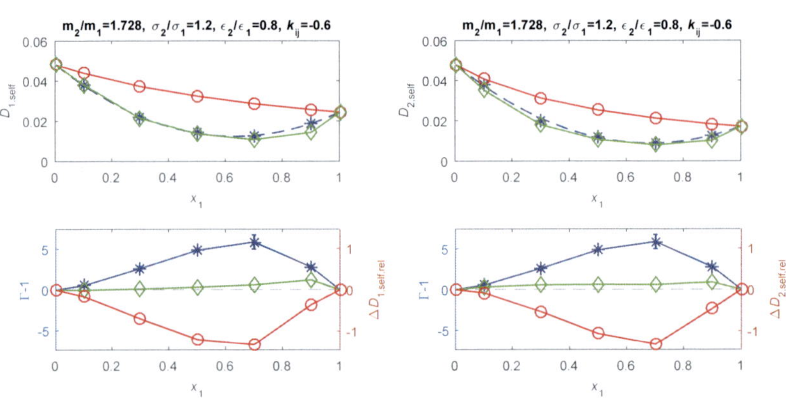
m_2/m_1=1.728, σ_2/σ_1=1.2, ϵ_2/ϵ_1=0.8, k_{ij}=-0.6
$D_{1,self}$
$D_{2,self}$
x_1
Γ-1
$\Delta D_{1,self,rel}$
$\Delta D_{2,self,rel}$

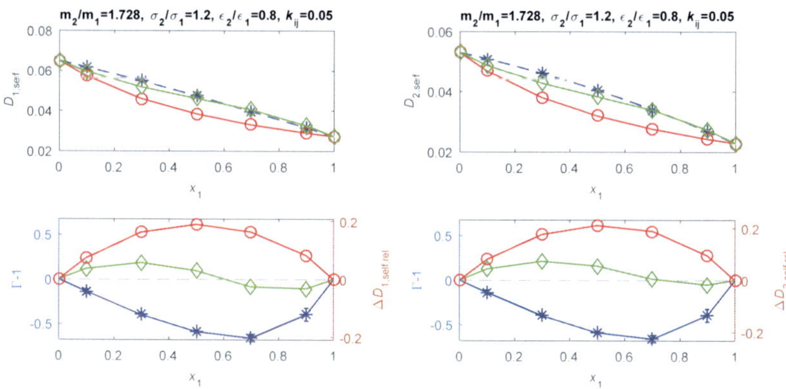
m2/m1=1.728, σ2/σ1=1.2, ε2/ε1=0.8, kij=0.05
m2/m1=1.728, σ2/σ1=1.2, ε2/ε1=0.8, kij=0.05
D1,self
D2,self
x1
Γ-1
ΔD1,self,rel
ΔD2,self,rel

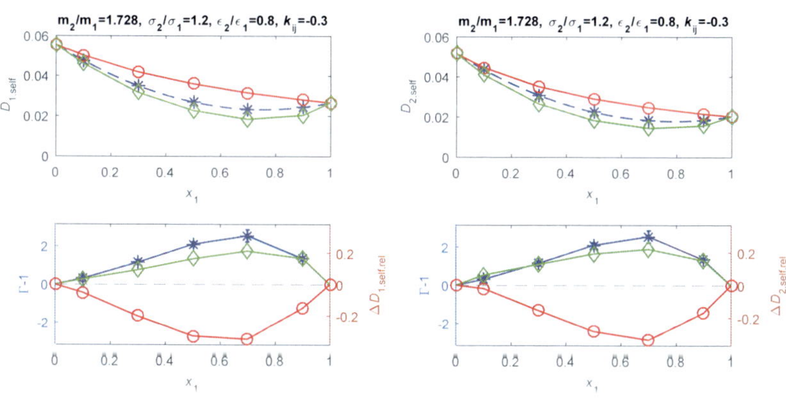
m2/m1=1.728, σ2/σ1=1.2, ε2/ε1=0.8, kij=-0.3
m2/m1=1.728, σ2/σ1=1.2, ε2/ε1=0.8, kij=-0.3
D1,self
D2,self
x1
Γ-1
ΔD1,self,rel
ΔD2,self,rel

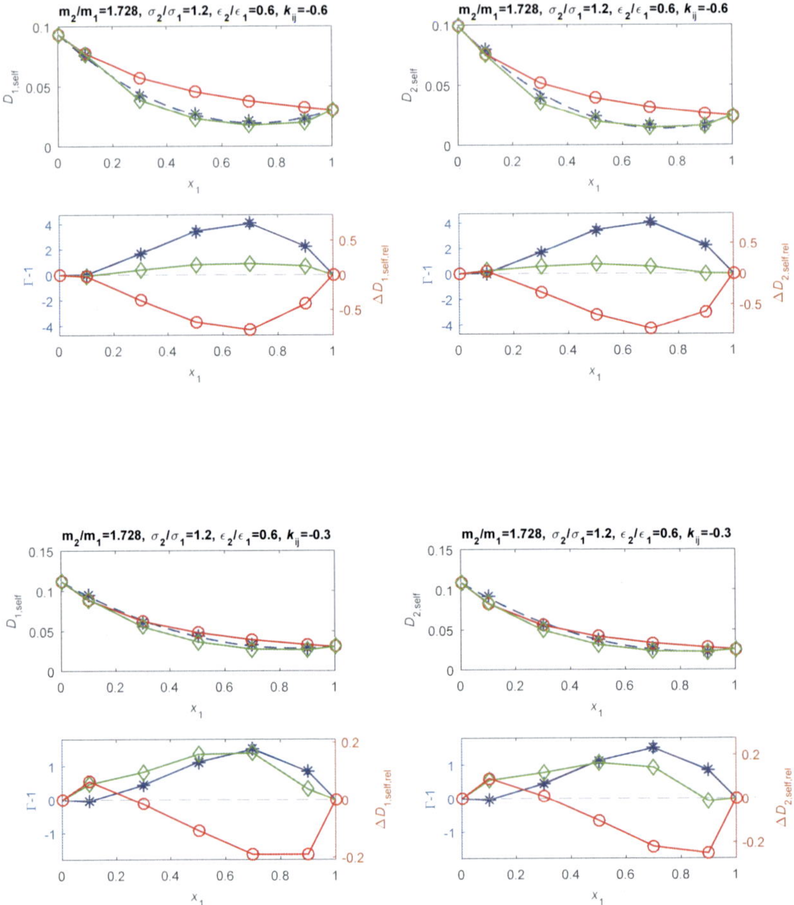

m_2/m_1=1.728, σ_2/σ_1=1.2, ε_2/ε_1=0.6, k_ij=-0.6
D_1,self
D_2,self
x_1
Γ-1
ΔD_1,self,rel
ΔD_2,self,rel
m_2/m_1=1.728, σ_2/σ_1=1.2, ε_2/ε_1=0.6, k_ij=-0.3

A.3.3 LJ systems with molar mass ratios $m_2/m_1 > 2$

Composition-dependent self-diffusion coefficients $D_{i,\text{self}}$, thermodynamic factors $\Gamma - 1$, and relative deviations $\Delta D_{i,\text{self,rel}}$ of LJ systems with molar mass ratios $m_2/m_1 > 2$. The specifications of the LJ systems ϵ_2/ϵ_1, σ_2/σ_1, m_2/m_1, and k_{ij} are given in the title of each figure.

Top figures: Blue stars: Simulation results of self-diffusion coefficients $D_{i,\text{self}}$ of binary LJ systems as function of the mole fraction x_1 of the first species. Blue dashed line: smoothing fit to the simulation results; red circles/line: predictions of the McCarty-Mason equation (Equation 6); green diamonds/line: predictions of the modified McCarty-Mason equation (Equation 25). The error bars of $D_{i,\text{self}}$ are smaller than the symbols in most cases. Please note that y-axes are adapted for each system.

Bottom figures: Composition dependence of the thermodynamic factor $\Gamma - 1$ (blue stars/line, left axis) and composition dependence of the relative deviation $\Delta D_{i,\text{self,rel}}$ between the self-diffusion coefficients and the predictions of the McCarty-Mason equation (Equation 6) (red circles/line, right axis) and the modified McCarthy-Mason equation (Equation 25) (green diamonds/line, right axis). A clear correlation between $\Gamma - 1$ and $\Delta D_{i,\text{self,rel}}$ can be observed. The error bars of $\Gamma - 1$ are smaller than the symbols in most cases. Please note that y-axes are adapted for each system.

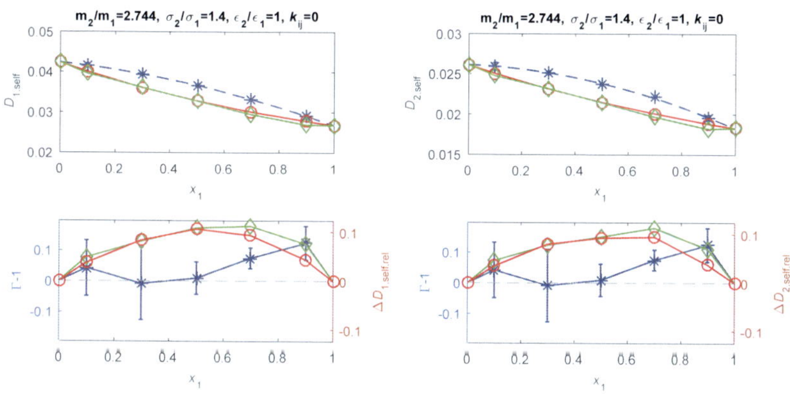

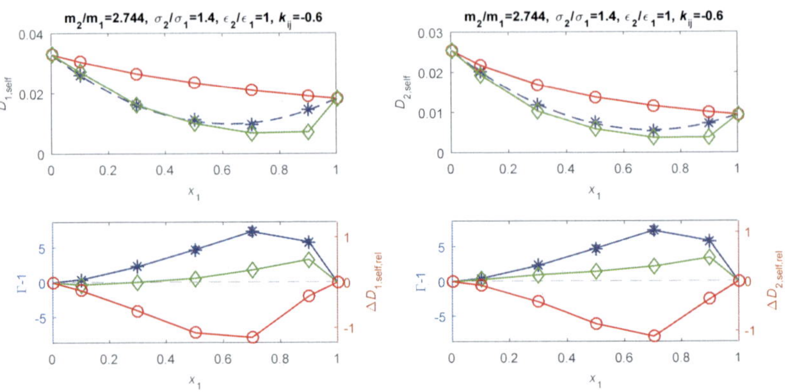

m_2/m_1=2.744, σ_2/σ_1=1.4, ϵ_2/ϵ_1=1, k_{ij}=-0.6
$D_{1,self}$
$D_{2,self}$
x_1
Γ-1
$\Delta D_{1,self,rel}$
$\Delta D_{2,self,rel}$

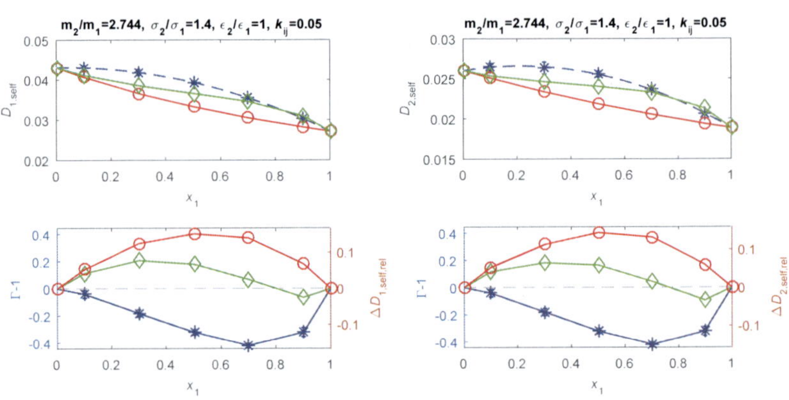

m_2/m_1=2.744, σ_2/σ_1=1.4, ϵ_2/ϵ_1=1, k_{ij}=0.05
$D_{1,self}$
$D_{2,self}$
x_1
Γ-1
$\Delta D_{1,self,rel}$
$\Delta D_{2,self,rel}$

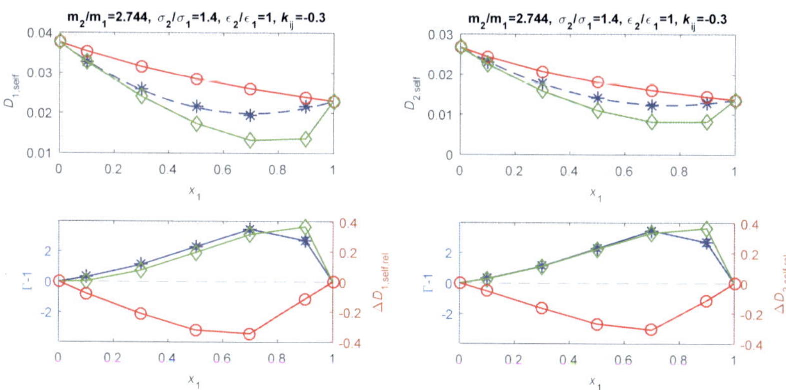
m2/m1=2.744, σ2/σ1=1.4, ε2/ε1=1, kij=-0.3
D1,self
x1
Γ-1
ΔD1,self,rel
m2/m1=2.744, σ2/σ1=1.4, ε2/ε1=1, kij=-0.3
D2,self
x1
Γ-1
ΔD2,self,rel

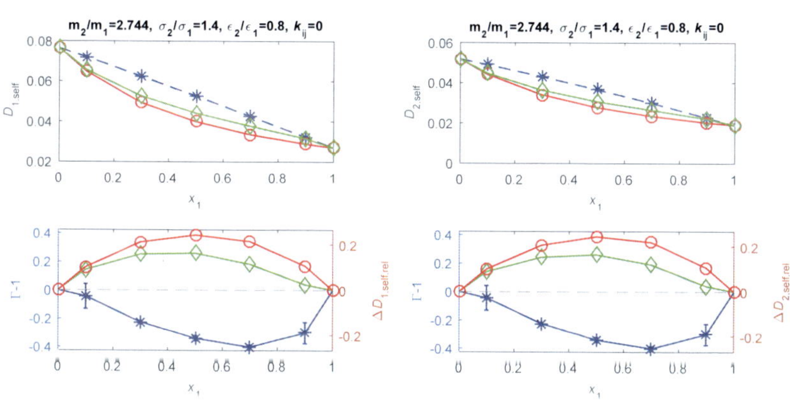
m2/m1=2.744, σ2/σ1=1.4, ε2/ε1=0.8, kij=0
D1,self
x1
Γ-1
ΔD1,self,rel
m2/m1=2.744, σ2/σ1=1.4, ε2/ε1=0.8, kij=0
D2,self
x1
Γ-1
ΔD2,self,rel

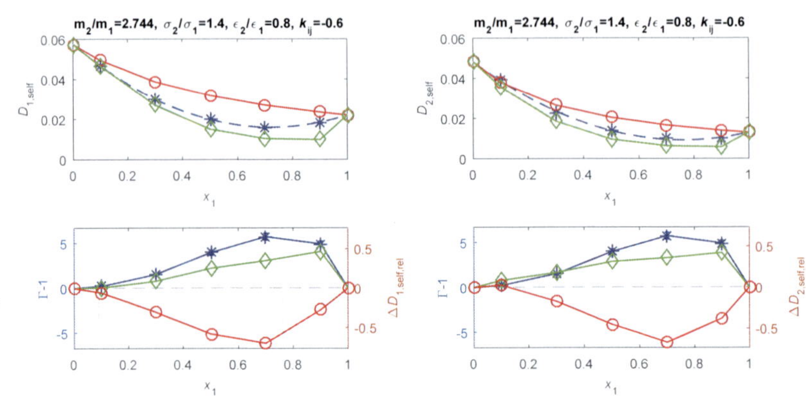

m$_2$/m$_1$=2.744, σ_2/σ_1=1.4, ϵ_2/ϵ_1=0.8, k_{ij}=-0.6
$D_{1,\text{self}}$
x_1
Γ-1
$\Delta D_{1,\text{self,rel}}$
m$_2$/m$_1$=2.744, σ_2/σ_1=1.4, ϵ_2/ϵ_1=0.8, k_{ij}=-0.6
$D_{2,\text{self}}$
x_1
Γ-1
$\Delta D_{2,\text{self,rel}}$

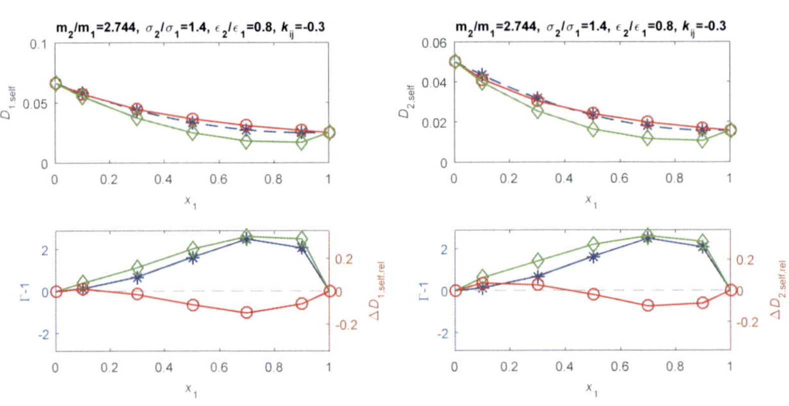

m$_2$/m$_1$=2.744, σ_2/σ_1=1.4, ϵ_2/ϵ_1=0.8, k_{ij}=-0.3
$D_{1,\text{self}}$
x_1
Γ-1
$\Delta D_{1,\text{self,rel}}$
m$_2$/m$_1$=2.744, σ_2/σ_1=1.4, ϵ_2/ϵ_1=0.8, k_{ij}=-0.3
$D_{2,\text{self}}$
x_1
Γ-1
$\Delta D_{2,\text{self,rel}}$

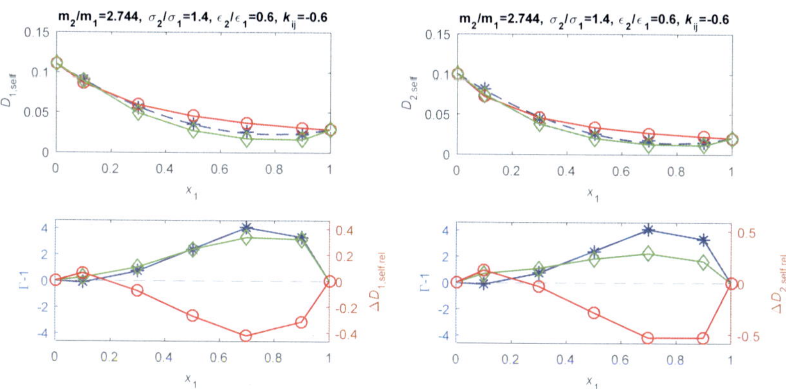

m$_2$/m$_1$=2.744, σ$_2$/σ$_1$=1.4, ε$_2$/ε$_1$=0.6, k$_{ij}$=-0.6
$D_{1,self}$
$D_{2,self}$
x_1
Γ-1
$\Delta D_{1,self,rel}$
$\Delta D_{2,self,rel}$

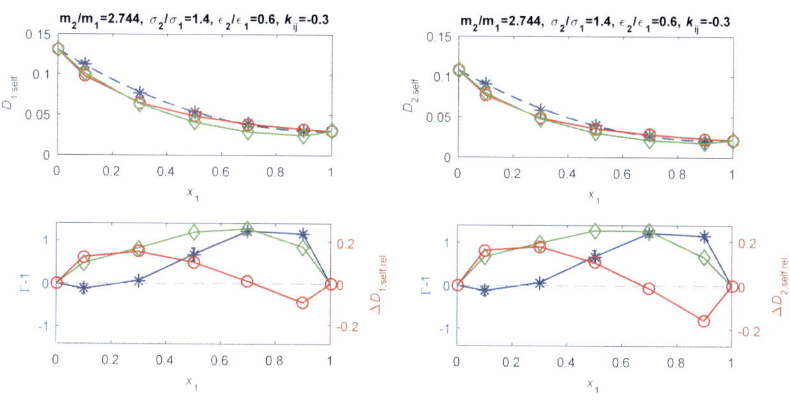

m$_2$/m$_1$=2.744, σ$_2$/σ$_1$=1.4, ε$_2$/ε$_1$=0.6, k$_{ij}$=-0.3
$D_{1,self}$
$D_{2,self}$
x_1
Γ-1
$\Delta D_{1,self,rel}$
$\Delta D_{2,self,rel}$

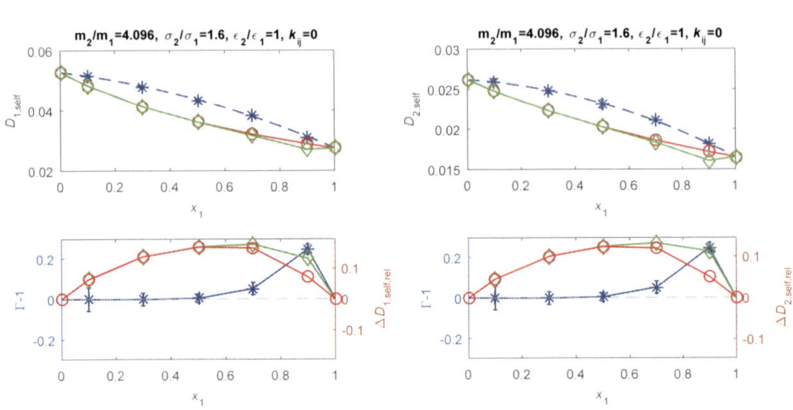

m_2/m_1=4.096, σ_2/σ_1=1.6, ϵ_2/ϵ_1=1, k_ij=0
D_1,self
D_2,self
x_1
Γ-1
ΔD_1,self,rel
ΔD_2,self,rel

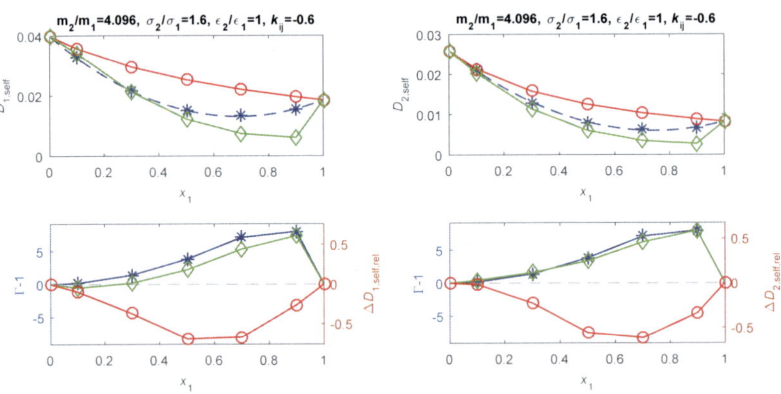

m_2/m_1=4.096, σ_2/σ_1=1.6, ϵ_2/ϵ_1=1, k_ij=-0.6
D_1,self
D_2,self
x_1
Γ-1
ΔD_1,self,rel
ΔD_2,self,rel

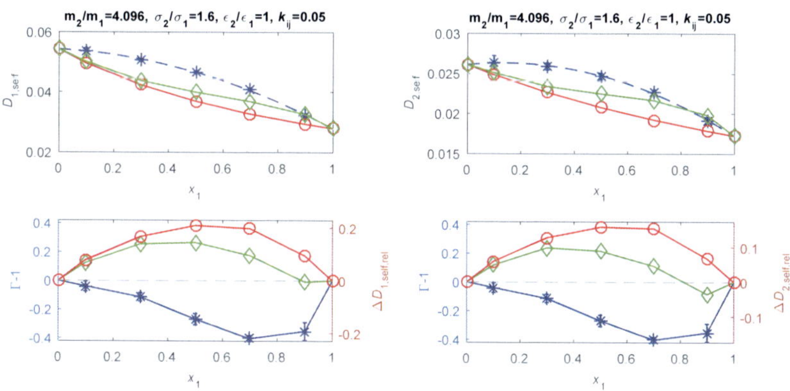
m2/m1=4.096, σ2/σ1=1.6, ε2/ε1=1, kij=0.05
D1,self
x1
Γ-1
ΔD1,self,rel
m2/m1=4.096, σ2/σ1=1.6, ε2/ε1=1, kij=0.05
D2,self
x1
Γ-1
ΔD2,self,rel

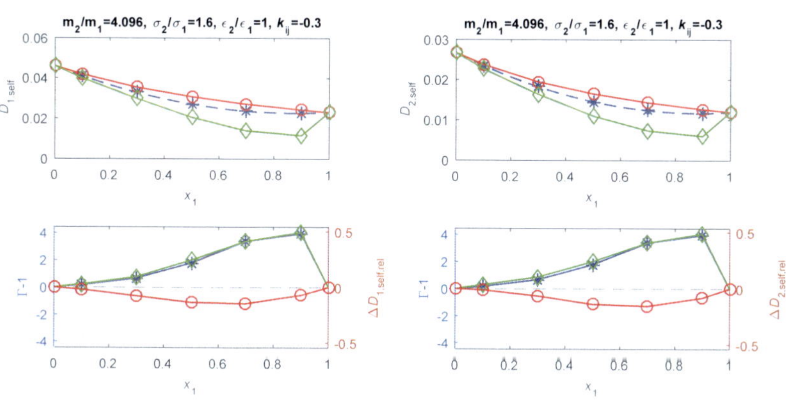
m2/m1=4.096, σ2/σ1=1.6, ε2/ε1=1, kij=-0.3
D1,self
x1
Γ-1
ΔD1,self,rel
m2/m1=4.096, σ2/σ1=1.6, ε2/ε1=1, kij=-0.3
D2,self
x1
Γ-1
ΔD2,self,rel

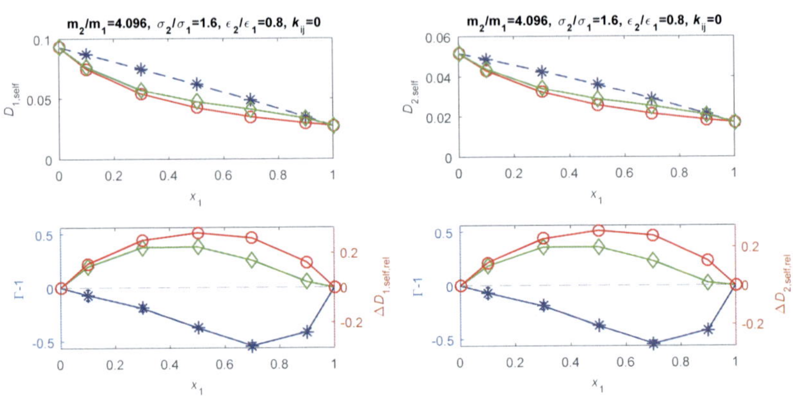
m2/m1=4.096, σ2/σ1=1.6, ϵ2/ϵ1=0.8, kij=0
D1,self
x1
Γ-1
ΔD1,self,rel
m2/m1=4.096, σ2/σ1=1.6, ϵ2/ϵ1=0.8, kij=0
D2,self
x1
Γ-1
ΔD2,self,rel

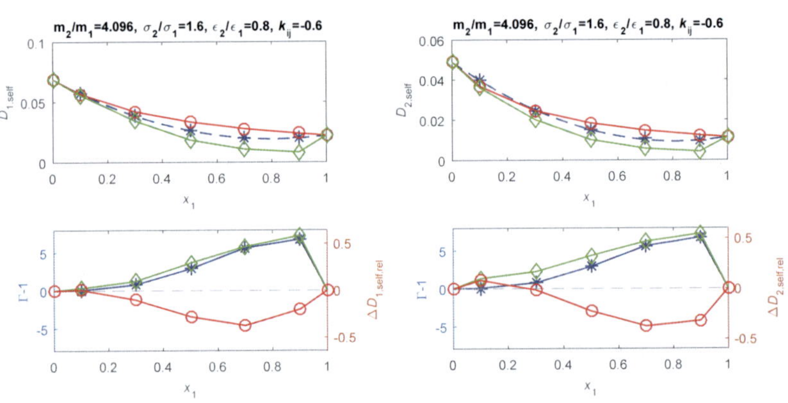
m2/m1=4.096, σ2/σ1=1.6, ϵ2/ϵ1=0.8, kij=-0.6
D1,self
x1
Γ-1
ΔD1,self,rel
m2/m1=4.096, σ2/σ1=1.6, ϵ2/ϵ1=0.8, kij=-0.6
D2,self
x1
Γ-1
ΔD2,self,rel

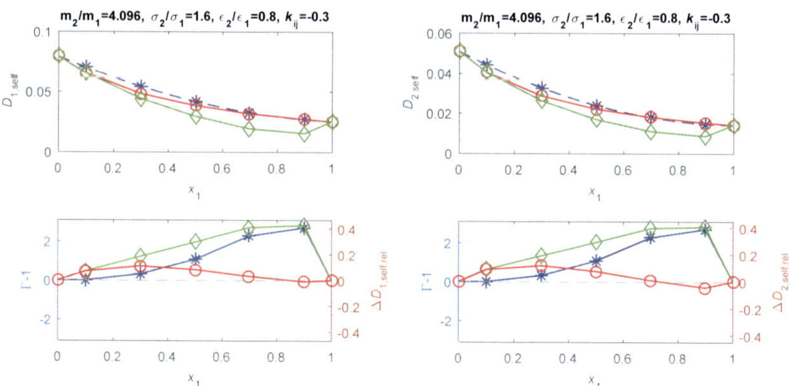

m2/m1=4.096, σ2/σ1=1.6, ϵ2/ϵ1=0.8, kij=-0.3
D1,self
D2,self
x1
Γ-1
ΔD1,self,rel
ΔD2,self,rel

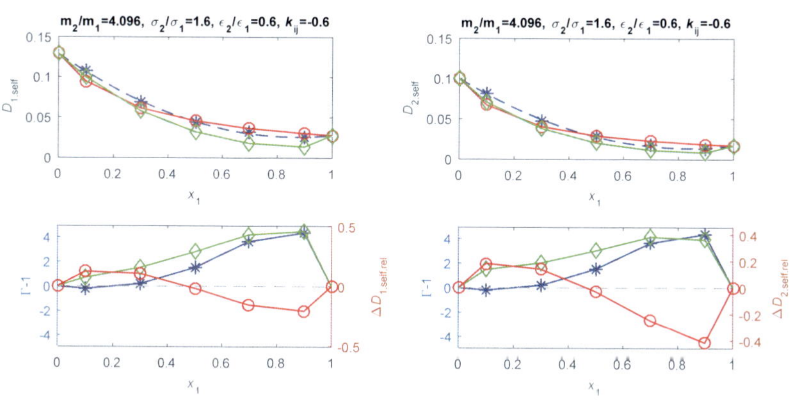

m2/m1=4.096, σ2/σ1=1.6, ϵ2/ϵ1=0.6, kij=-0.6
D1,self
D2,self
x1
Γ-1
ΔD1,self,rel
ΔD2,self,rel

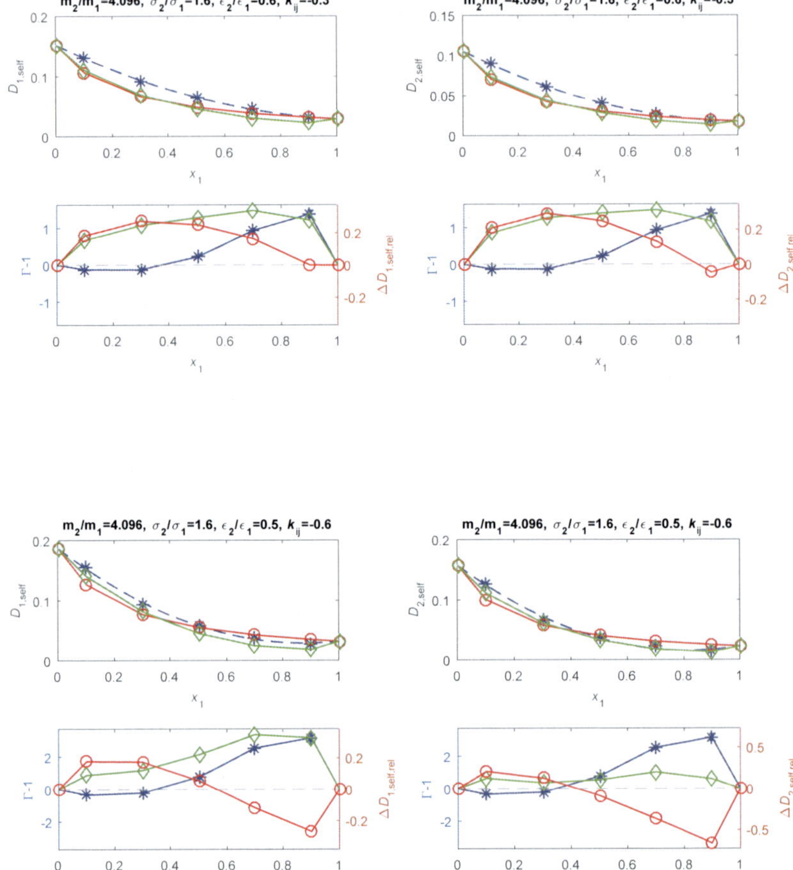

m_2/m_1=4.096, σ_2/σ_1=1.6, ε_2/ε_1=0.6, k_ij=-0.3
m_2/m_1=4.096, σ_2/σ_1=1.6, ε_2/ε_1=0.5, k_ij=-0.6
$D_{1,self}$
$D_{2,self}$
x_1
Γ-1
$\Delta D_{1,self,rel}$
$\Delta D_{2,self,rel}$

A.4 Self-diffusion coefficients of molecular systems (experimental data)

A.4.1 Relative deviations $\Delta D_{2,\mathrm{self,rel}}$ of the McCarty-Mason prediction as a function of the thermodynamic factor Γ for component 2

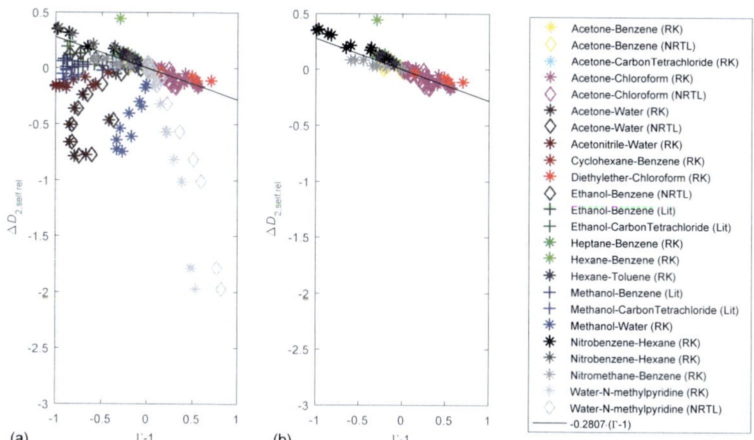

Figure A.2: Relative deviations $\Delta D_{2,\mathrm{self,rel}}$ of the McCarty-Mason prediction (Equation (6)) as function of the thermodynamic factor Γ for molecular systems (symbols) and linear fit of $\Delta D_{2,\mathrm{self,rel}}$ derived from LJ systems (black line, cf. Equation (23)). Stars: Experimental data with thermodynamic factors calculated with Redlich-Kister (RK). Diamonds: Experimental data with thermodynamic factors calculated with NRTL. Plus symbols: Experimental data with thermodynamic factors reported in literature.
(a) $\Delta D_{2,\mathrm{self,rel}}$ for all considered molecular systems.
(b) $\Delta D_{2,\mathrm{self,rel}}$ for molecular systems with molar mass ratios $M_2/M_1 < 2$ and without dimerizing species.

A.4.2 Molecular systems with molar mass ratios $M_2/M_1 < 2$ and without dimerizing species

Composition-dependent self-diffusion coefficients $D_{i,\text{self}}$, thermodynamic factors $\Gamma - 1$, and relative deviations $\Delta D_{i,\text{self,rel}}$ of molecular systems with molar mass ratios $M_2/M_1 < 2$ and without dimerizing species. The specific components of each molecular system are given in the title of each figure.

Top figures: Blue stars: Experimental data of composition-dependent self-diffusion coefficients $D_{i,\text{self}}$. Blue dashed line: smoothing fit of the experimental self-diffusion coefficients; red circles/line: predictions of the McCarty-Mason equation (Equation 6); green diamonds/line: predictions of the modified McCarty-Mason equation (Equation 25).

Bottom figures: Composition dependence of the thermodynamic factor $\Gamma - 1$ (blue symbols/line, left axis) and composition dependence of the relative deviation $\Delta D_{i,\text{self,rel}}$ between the experimental self-diffusion coefficients and the predictions of the McCarty-Mason equation (Equation 6) (red circles/line, right axis) and the modified McCarthy-Mason equation (Equation 25) (green symbols/line, right axis). The symbols for the thermodynamic factors and the predictions of the modified McCarty-Mason predictions indicate the source of the thermodynamic factor calculations: Redlich-Kister (stars), NRTL (diamonds), or reported directly in literature (crosses).

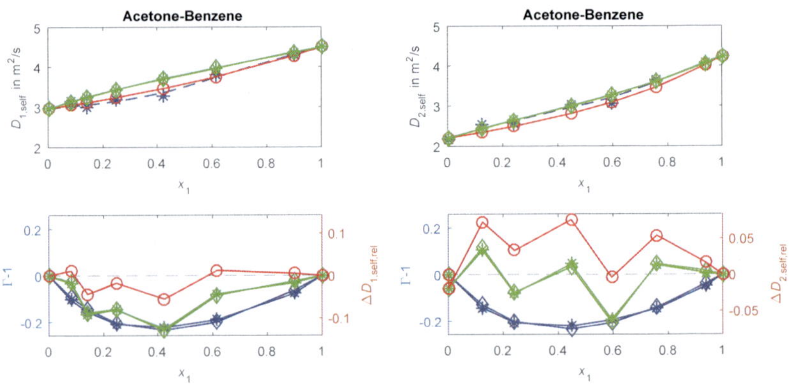

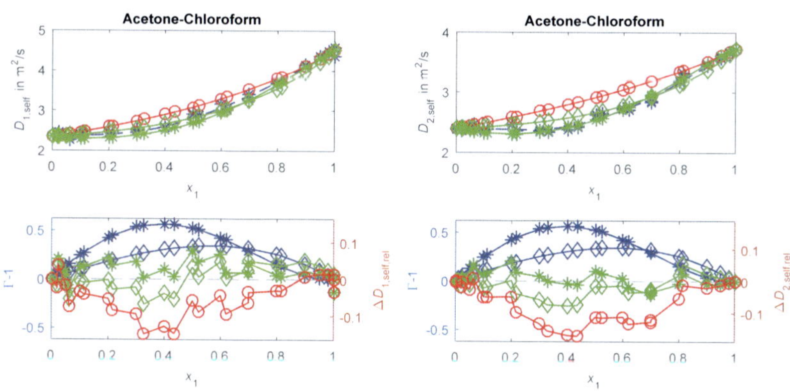

Acetone-Chloroform
$D_{1,self}$ in m^2/s
x_1
Γ-1
$\Delta D_{1,self,rel}$
Acetone-Chloroform
$D_{2,self}$ in m^2/s
x_1
Γ-1
$\Delta D_{2,self,rel}$

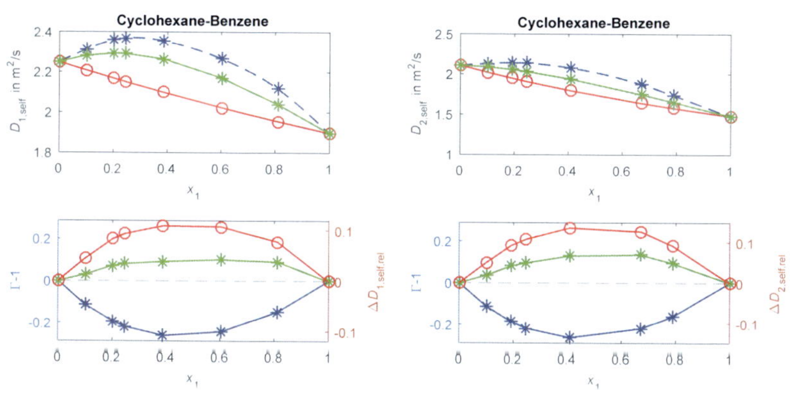

Cyclohexane-Benzene
$D_{1,self}$ in m^2/s
x_1
Γ-1
$\Delta D_{1,self,rel}$
Cyclohexane-Benzene
$D_{2,self}$ in m^2/s
x_1
Γ-1
$\Delta D_{2,self,rel}$

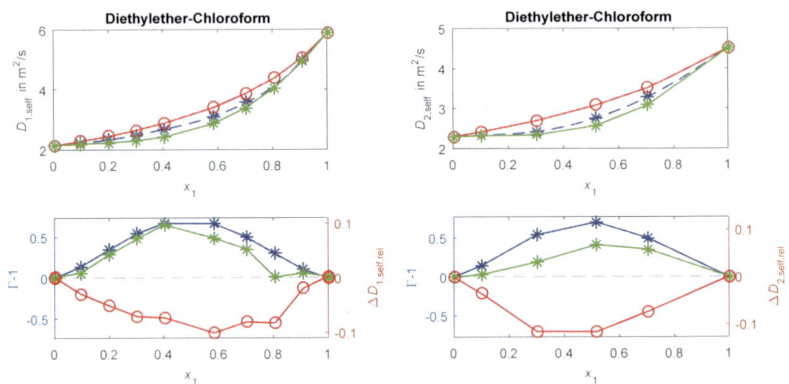
Diethylether-Chloroform
$D_{1,self}$ in m^2/s
x_1
Diethylether-Chloroform
$D_{2,self}$ in m^2/s
x_1
Γ-1
$\Delta D_{1,self,rel}$
x_1
Γ-1
$\Delta D_{2,self,rel}$
x_1

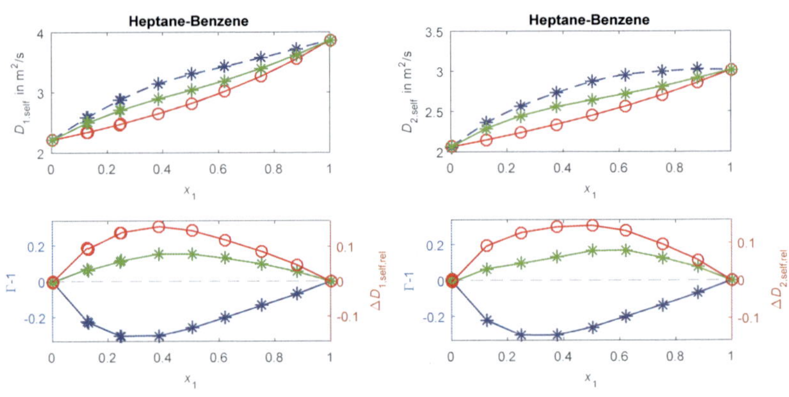
Heptane-Benzene
$D_{1,self}$ in m^2/s
x_1
Heptane-Benzene
$D_{2,self}$ in m^2/s
x_1
Γ-1
$\Delta D_{1,self,rel}$
x_1
Γ-1
$\Delta D_{2,self,rel}$
x_1

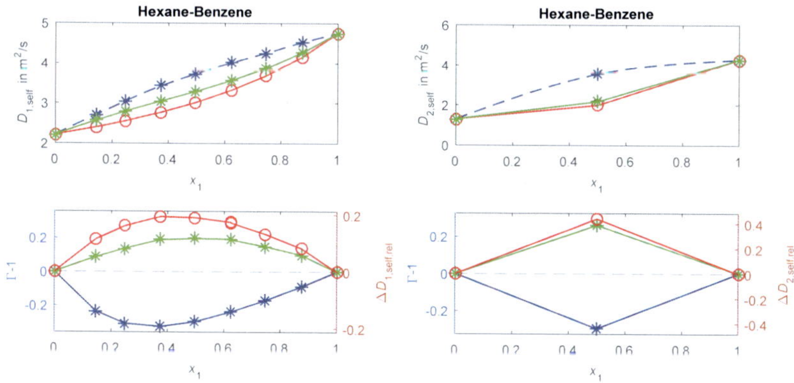
Hexane-Benzene
$D_{1,self}$ in m^2/s
x_1
Γ-1
$\Delta D_{1,self,rel}$
Hexane-Benzene
$D_{2,self}$ in m^2/s
x_1
Γ-1
$\Delta D_{2,self,rel}$

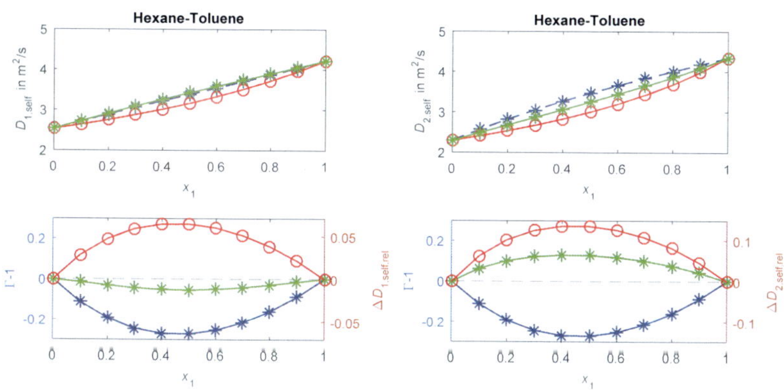
Hexane-Toluene
$D_{1,self}$ in m^2/s
x_1
Γ-1
$\Delta D_{1,self,rel}$
Hexane-Toluene
$D_{2,self}$ in m^2/s
x_1
Γ-1
$\Delta D_{2,self,rel}$

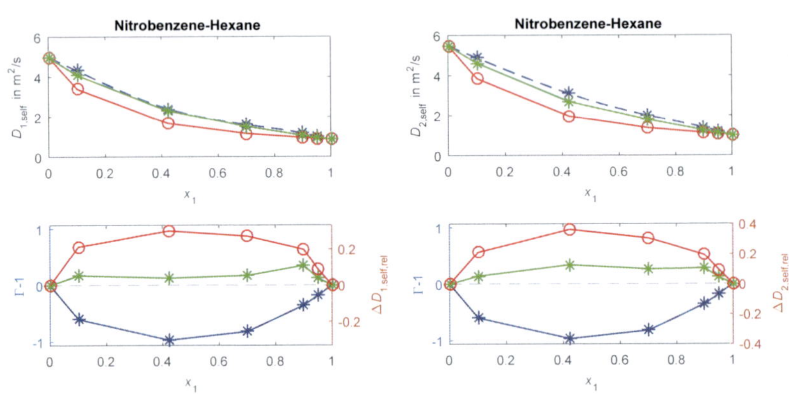
Nitrobenzene-Hexane
$D_{1,self}$ in m^2/s
x_1
Nitrobenzene-Hexane
$D_{2,self}$ in m^2/s
x_1
Γ-1
$\Delta D_{1,self,rel}$
x_1
Γ-1
$\Delta D_{2,self,rel}$
x_1

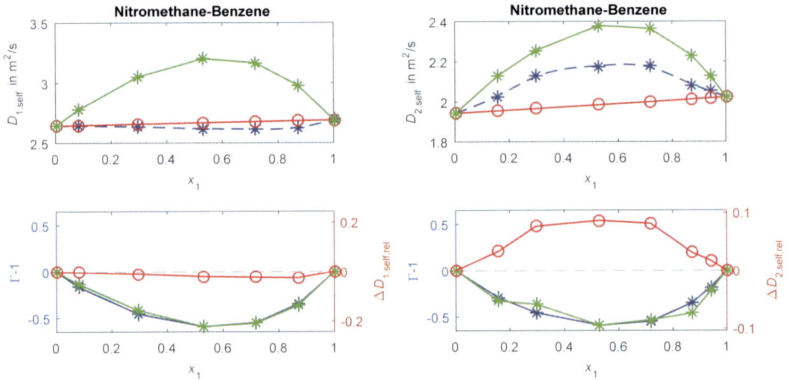
Nitromethane-Benzene
$D_{1,self}$ in m^2/s
x_1
Nitromethane-Benzene
$D_{2,self}$ in m^2/s
x_1
Γ-1
$\Delta D_{1,self,rel}$
x_1
Γ-1
$\Delta D_{2,self,rel}$
x_1

A.4.3 Molecular systems with molar mass ratios $M_2/M_1 > 2$ and/or with dimerizing species

Composition-dependent self-diffusion coefficients $D_{i,\text{self}}$, thermodynamic factors $\Gamma - 1$, and relative deviations $\Delta D_{i,\text{self,rel}}$ of molecular systems with molar mass ratios $M_2/M_1 > 2$ and/or with dimerizing species. The specific components of each molecular system are given in the title of each figure.

Top figures: Blue stars: Experimental data of composition-dependent self-diffusion coefficients $D_{i,\text{self}}$. Blue dashed line: smoothing fit of the experimental self-diffusion coefficients; red circles/line: predictions of the McCarty-Mason equation (Equation 6); green diamonds/line: predictions of the modified McCarty-Mason equation (Equation 25).

Bottom figures: Composition dependence of the thermodynamic factor $\Gamma - 1$ (blue symbols/line, left axis) and composition dependence of the relative deviation $\Delta D_{i,\text{self,rel}}$ between the experimental self-diffusion coefficients and the predictions of the McCarty-Mason equation (Equation 6) (red circles/line, right axis) and the modified McCarthy-Mason equation (Equation 25) (green symbols/line, right axis). The symbols for the thermodynamic factors and the predictions of the modified McCarty-Mason predictions indicate the source of the thermodynamic factor calculations: Redlich-Kister (stars), NRTL (diamonds), or reported directly in literature (crosses).

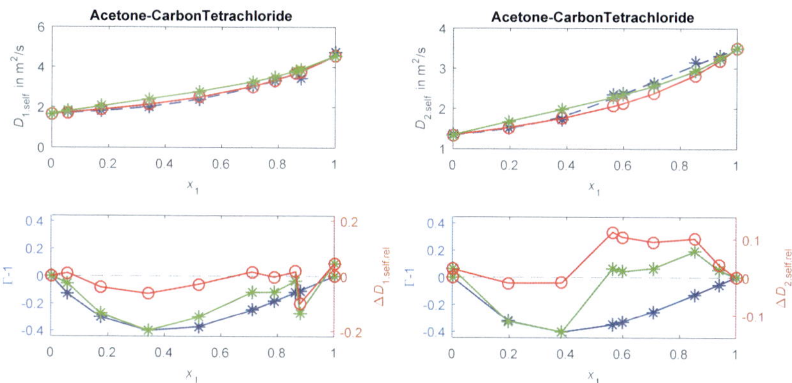

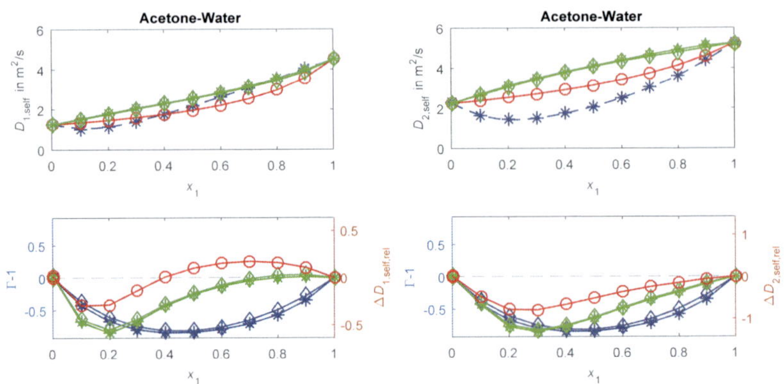
Acetone-Water
$D_{1,self}$ in m^2/s
x_1
Γ-1
$\Delta D_{1,self,rel}$
Acetone-Water
$D_{2,self}$ in m^2/s
x_1
Γ-1
$\Delta D_{2,self,rel}$

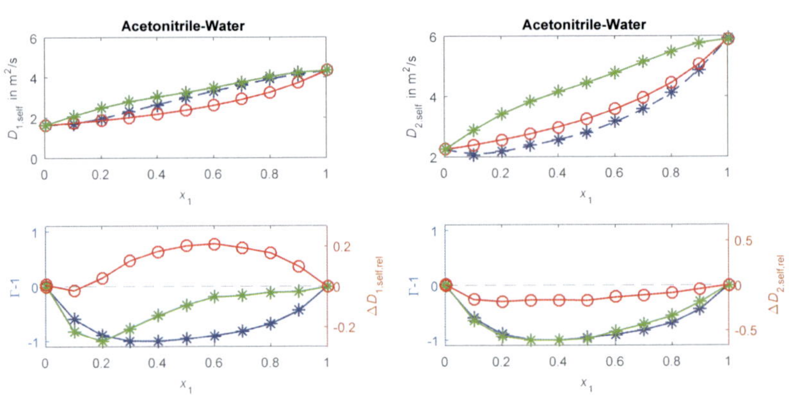
Acetonitrile-Water
$D_{1,self}$ in m^2/s
x_1
Γ-1
$\Delta D_{1,self,rel}$
Acetonitrile-Water
$D_{2,self}$ in m^2/s
x_1
Γ-1
$\Delta D_{2,self,rel}$

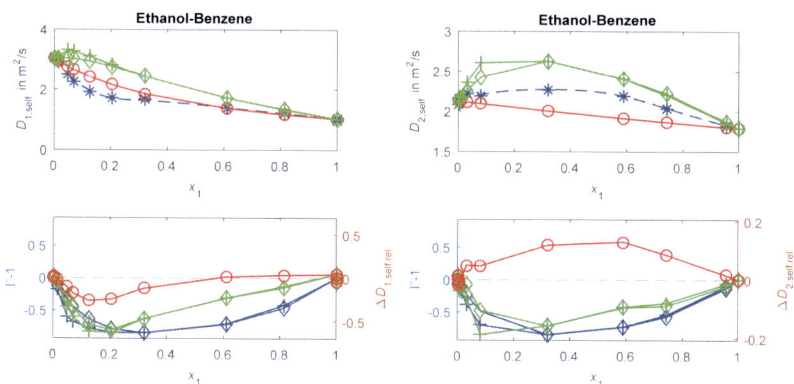
Ethanol-Benzene
$D_{1,self}$ in m^2/s
x_1
Γ-1
$\Delta D_{1,self,rel}$
Ethanol-Benzene
$D_{2,self}$ in m^2/s
x_1
Γ-1
$\Delta D_{2,self,rel}$

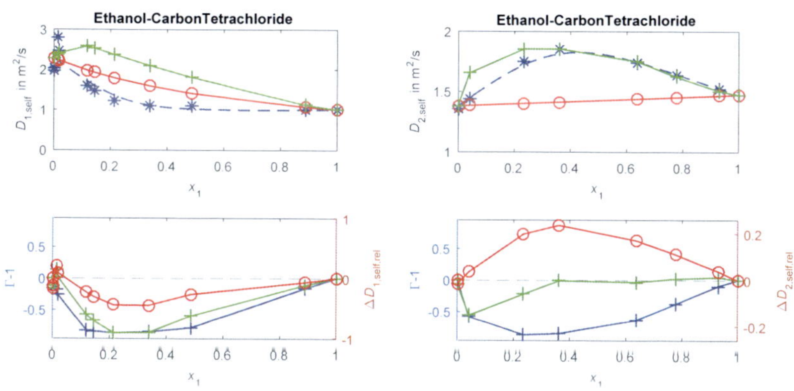
Ethanol-CarbonTetrachloride
$D_{1,self}$ in m^2/s
x_1
Γ-1
$\Delta D_{1,self,rel}$
Ethanol-CarbonTetrachloride
$D_{2,self}$ in m^2/s
x_1
Γ-1
$\Delta D_{2,self,rel}$

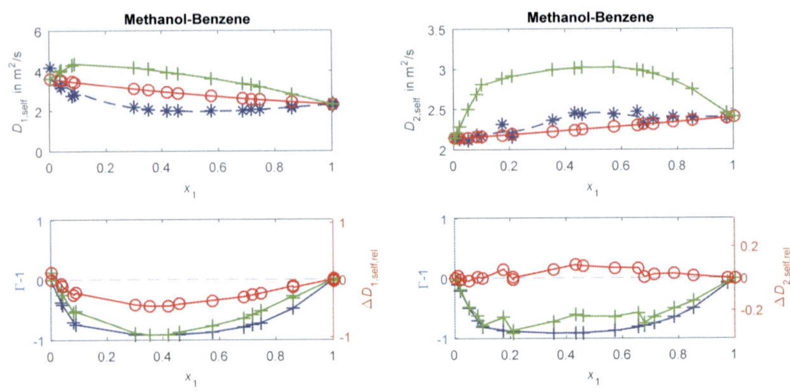
Methanol-Benzene
$D_{1,self}$ in m²/s
x_1
Methanol-Benzene
$D_{2,self}$ in m²/s
x_1
Γ-1
$\Delta D_{1,self,rel}$
x_1
Γ-1
$\Delta D_{2,self,rel}$
x_1

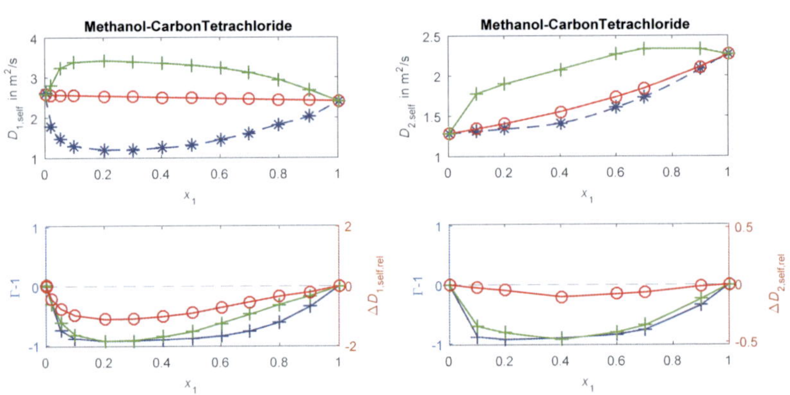
Methanol-CarbonTetrachloride
$D_{1,self}$ in m²/s
x_1
Methanol-CarbonTetrachloride
$D_{2,self}$ in m²/s
x_1
Γ-1
$\Delta D_{1,self,rel}$
x_1
Γ-1
$\Delta D_{2,self,rel}$
x_1

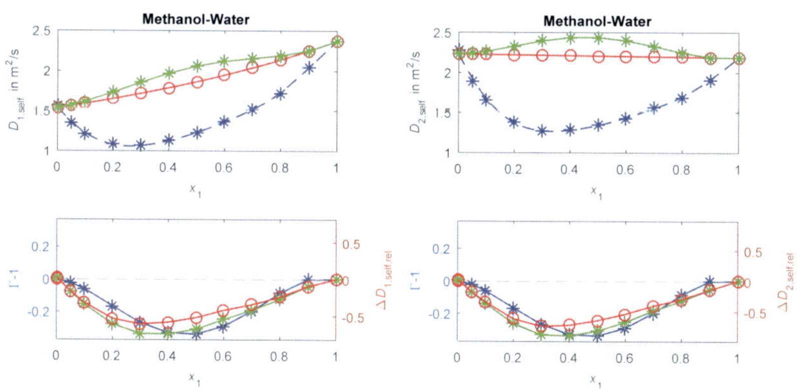
Methanol-Water
$D_{1,self}$ in m²/s
x_1
Methanol-Water
$D_{2,self}$ in m²/s
x_1
Γ-1
$\Delta D_{1,self,rel}$
Γ-1
$\Delta D_{2,self,rel}$

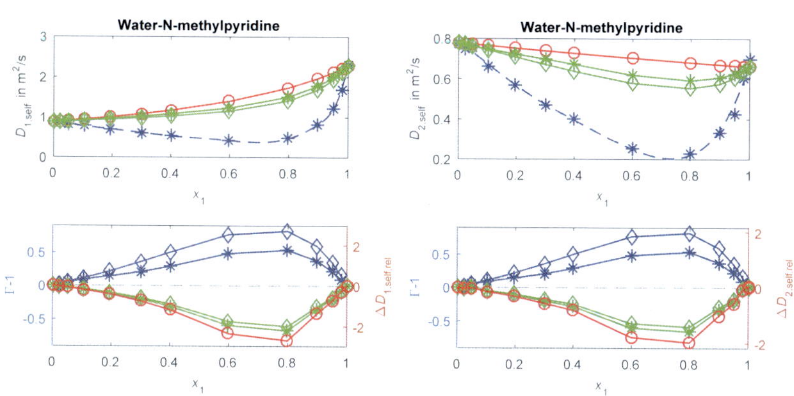
Water-N-methylpyridine
$D_{1,self}$ in m²/s
x_1
Water-N-methylpyridine
$D_{2,self}$ in m²/s
x_1
Γ-1
$\Delta D_{1,self,rel}$
Γ-1
$\Delta D_{2,self,rel}$

Appendix B

Model-based optimal experimental design of diffusion experiments: Additional data

B.1 Analytical solutions for the concentration profiles

The analytical solutions for the concentrations as a function of position and time are given below. They are taken from Carslaw and Jaeger (1959). J_0 and J_1 are the Bessel functions of zero and first order, respectively. The model for the velocity profile in the microfluidic channel is taken from (Rohsenow et al., 1998).

Free diffusion

$$\bar{c} = \frac{1}{2}\left[1 - \operatorname{erf}\left\{\frac{\xi - \xi_0}{2\sqrt{Fo}}\right\}\right] \tag{B.1}$$

Closed cell

$$\bar{c} = \xi_0 + \frac{2}{\pi}\sum_{n=1}^{\infty}\frac{1}{n}\sin\left(n\pi\xi_0\right)\exp\left(-n^2\pi^2 Fo\right)\cos\left(n\pi\xi\right) \tag{B.2}$$

Microfluidics

$$\frac{v(\xi)}{\bar{v}} \cdot \frac{\partial \bar{c}(\xi, Fo)}{\partial Fo} = D\frac{\partial^2 \bar{c}(\xi, Fo)}{\partial \xi^2}. \tag{B.3}$$

Here, $v(\xi)$ is the velocity profile averaged over the height of the microfluidic channel. v is the average velocity in the channel. Thus, $v(\xi)/v$ is the dimensionless velocity profile which is given by (Rohsenow et al., 1998)

$$\frac{v(\xi)}{\bar{v}} = \left(\frac{m+1}{m}\right)[1-(2\xi)^m] \quad \text{(B.4)}$$

$$\text{with } m = 1.7 + 0.5\left(\frac{H}{L}\right)^{-1.4}.$$

H/L is the ratio of channel height to channel width which is assumed to be $H/L = 0.1$.

The initial and boundary conditions are given by

$$\frac{\partial \bar{c}(\xi, Fo=0)}{\partial Fo} = \begin{cases} 1 & \text{for } \xi < \xi_0 \\ 0 & \text{for } \xi \geq \xi_0 \end{cases}, \quad \text{(B.5)}$$

$$\frac{\partial \bar{c}(\xi=0, Fo)}{\partial Fo} = \frac{\partial \bar{c}(\xi=1, Fo)}{\partial Fo} = 0. \quad \text{(B.6)}$$

Closed cylinder

$$\bar{c} = \xi_0^2 + 2\sum_{n=1}^{\infty} \exp\left(-\beta_n^2 Fo\right) \frac{J_0(\beta_n \xi) \cdot J_1(\beta_n \xi_0)\,\xi_0}{\beta_n \cdot J_0^2(\beta_n)} \quad \text{(B.7)}$$

where β_n are the roots of $J_1(\beta_n) = 0$.

Closed sphere

$$\bar{c} = \xi_0^3 + \frac{2}{\xi}\sum_{n=1}^{\infty} \exp\left(-\beta_n^2 Fo\right) \frac{\sin(\beta_n \xi)}{\beta_n^2 \sin^2(\beta_n)} \left[\sin(\beta_n \xi_0) - \beta_n \xi_0 \cos(\beta_n \xi_0)\right] \quad \text{(B.8)}$$

where β_n are the roots of $\beta_n \cdot \cot(\beta_n) - 1 = 0$.

Semi-free diffusion

$$\bar{c} = -\operatorname{erf}\left\{\frac{\xi - 1}{2\sqrt{Fo}}\right\} \quad \text{(B.9)}$$

Open cell

$$\bar{c} = \frac{4}{\pi}\sum_{n=0}^{\infty} \frac{(-1)^n}{2n+1} \exp\left(-\frac{(2n+1)^2\pi^2}{4} Fo\right) \cos\left(\frac{(2n+1)\pi}{2}\xi\right) \quad \text{(B.10)}$$

Open cylinder

$$\bar{c} = 2\sum_{n=1}^{\infty} \exp\left(-\beta_n^2 Fo\right) \frac{J_0\left(\beta_n \xi\right)}{\beta_n \cdot J_1\left(\beta_n\right)} \tag{B.11}$$

$$\bar{c}_{\text{av}} = \frac{8}{\pi^2} \sum_{n=0}^{\infty} \frac{1}{\beta_n^2} \exp\left(-\beta_n^2 Fo\right) \tag{B.12}$$

where β_n are the roots of $J_1(\beta_n) = 0$.

Open sphere

$$\bar{C} = -\frac{2}{\pi\xi} \sum_{n=1}^{\infty} \frac{(-1)^n}{n} \sin\left(n\pi\xi\right) \exp\left(-n^2\pi^2 Fo\right) \tag{B.13}$$

$$\bar{C}_{\text{av}} = \frac{6}{\pi^2} \sum_{n=1}^{\infty} \frac{1}{n^2} \exp\left(-n^2\pi^2 Fo\right) \tag{B.14}$$

B.2 Concentration profiles of all geometries

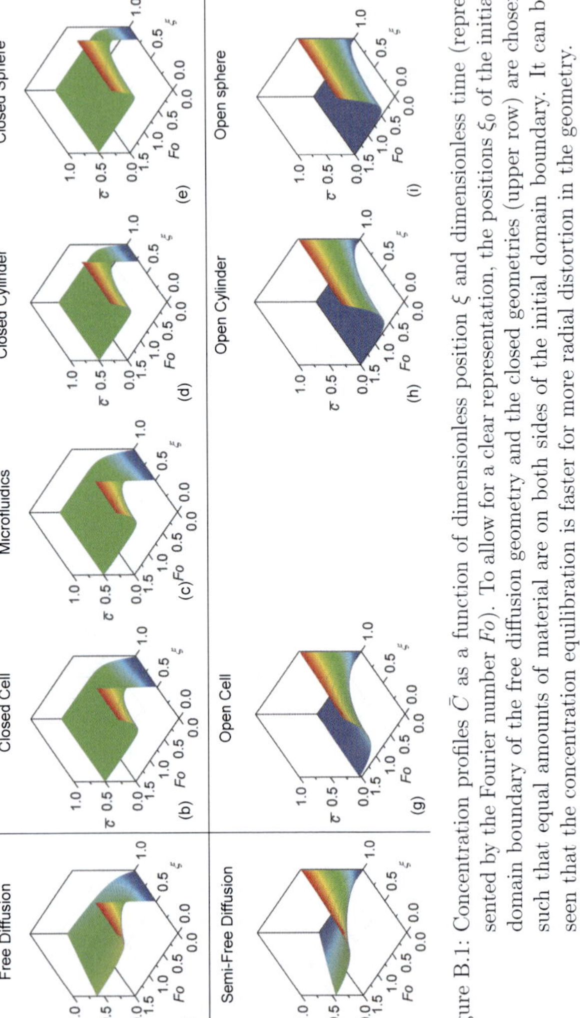

Figure B.1: Concentration profiles $\bar{C}$ as a function of dimensionless position ξ and dimensionless time (represented by the Fourier number Fo). To allow for a clear representation, the positions ξ_0 of the initial domain boundary of the free diffusion geometry and the closed geometries (upper row) are chosen such that equal amounts of material are on both sides of the initial domain boundary. It can be seen that the concentration equilibration is faster for more radial distortion in the geometry.

B.3 Examples for diffusion experiments in literature assigned to diffusion geometries, states of matter, and analytical techniques

Table B.1: Examples for diffusion experiments in literature assigned to diffusion geometries, states of matter, and analytical techniques

Geometry	Reference	Phase	Analytical technique
Free diffusion	Albright and Miller (1975)	liquid	Rayleigh interferometry
	Albright and Miller (1980)	liquid	Gouy interferometry
	Miller et al. (1996)	liquid	Gouy and Rayleigh interferometry
	Berg et al. (2007)	liquid	Raman spectroscopy
Closed cell	Leaist (1985)	liquid	Conductance measurements
	Wright et al. (1994)	liquid	Scintillation analysis
	Bardow et al. (2003, 2005a, 2006)	liquid	Raman spectroscopy
	Kugler et al. (2015b)	gas	Holographic interferometry
Microfluidics	Häusler et al. (2012)	liquid	UV-VIS
	Peters et al. (2017)	liquid	Raman spectroscopy
	Kamholz et al. (2001)	liquid	Fluorescence measurements
	Pappaert et al. (2005)	liquid	Fluorescence measurements
Closed cylinder	Helmke et al. (2004)	solid	Ion chromatrography, ion-selective electrodes
Closed Sphere	Li et al. (2015)	solid	X-Ray measurements
Open cell	Anderson and Saddington (1949)	liquid	Scintillation analysis
	Wang (1951)	liquid	Density measurements
	Miller et al. (1964)	liquid	Scintillation analysis
	Bacon and Adams (1970)	liquid	Fluorometric analysis
	Paccagnella et al. (1985)	solid	Electron microscopy
Open cylinder	Suárez et al. (2007)	solid	Current measurements
Open sphere	Tanaka et al. (1977)	solid	Pressure measurements
	Spiess et al. (2008)	liquid/solid	Confocal laser scanning microscopy
	Masuch (2011)	liquid/solid	Raman spectroscopy

B.4 Optimized experiments

Table B.2: Information in diffusion experiments as measured by the D-efficiencies ζ_D (Equation 3.23) for the optimal experiments for each geometry and all spatial and temporal resolutions. Due to different objective functions, only results within one column can be compared to each other.

Geometry	One position, one time	One position, over time	Spatially resolved, one time	Spatially resolved, over time	Average concentration, one time	Average concentration, over time
Free diffusion	0.03*	0.02*	0.03*	0.01*	-	-
Closed cell	0.14	0.07	0.13	0.06	-	-
Microfluidics	0.14	0.07	0.13	0.06	-	-
Closed cylinder	0.32	0.10	0.22	0.07	-	-
Closed sphere	0.50	0.12	0.35	0.08	-	-
Semi-free diffusion	0.14*	0.05*	0.14*	0.04*	-	-
Open cell	0.52	**1.00**	0.53	**1.00**	**1.00**	**1.00**
Open cylinder	0.76	0.61	0.76	0.61	0.75	0.34
Open sphere	**1.00**	0.46	**1.00**	0.47	0.64	0.17

*Free diffusion and semi-free diffusion equations are valid only as long as the walls have no or little influence on the diffusion process. The exact solution, considering the influence of the walls, is given by the equations of the closed cell and the open cell, respectively. In the optimization, a maximum error of 1% between concentrations in the free diffusion/semi-free diffusion geometry in comparison to the closed cell/open cell was allowed.

Table B.3: Optimal measurement positions ξ_{opt}, measurement times Fo_{opt}, and positions $\xi_{0,opt}$ of the initial domain boundary for different spatial and temporal resolutions.

Geometry		One position, one time	One position, over time	Spatially resolved, one time	Spatially resolved, over time	Average concentration, one time
Closed cell	ξ_{opt}	0	0	-	-	-
	Fo_{opt}	0.0413	-	0.1013	-	-
	$\xi_{0,opt}$	0.2844	0.4142	0.5000	0.5000	-
Microfluidics	ξ_{opt}	0	0	-	-	-
	Fo_{opt}	0.0413	-	0.1013	-	-
	$\xi_{0,opt}$	0.2844	0.4142	0.5000	0.5000	-
Closed cylinder	ξ_{opt}	0	0	-	-	-
	Fo_{opt}	0.0358	-	0.0730	-	-
	$\xi_{0,opt}$	0.3787	0.5391	0.5854	0.5944	-
Closed sphere	ξ_{opt}	0	0	-	-	-
	Fo_{opt}	0.0265	-	0.0590	-	-
	$\xi_{0,opt}$	0.3987	0.6114	0.6370	0.6524	-
Open cell	ξ_{opt}	0	0	-	-	-
	Fo_{opt}	0.4084	-	0.4053	-	0.4040
Open cylinder	ξ_{opt}	0	0	-	-	-
	Fo_{opt}	0.1965	-	0.1847	-	0.1600
Open sphere	ξ_{opt}	0	0	-	-	-
	Fo_{opt}	0.1340	-	0.1230	-	0.0800

B.5 Errors in the independent variables Fo, ξ, ξ_0

In real applications, not only errors in the concentration measurement $\bar{c}$ have to be expected, but errors in the setting of the independent variables, specified by the experimenter, have to be expected as well. By incorporating these errors in the design formulation, we can identify robust optimal designs which are insensitive to errors in the independent variables.

The independent variables specified by the experimenter are the measurement time, represented by the Fourier number Fo, the measurement position ξ, and the position ξ_0 of the initial domain boundary. If the independent variables contain errors, the local Fisher information M is given by (cf. Equation 3.9):

$$M = \left(\frac{\partial \bar{c}}{\partial D}\right)^2 \cdot \frac{1}{\sigma_{\bar{c}}^2 + \left(\frac{\partial \bar{c}}{\partial Fo}\right)^2 \sigma_{Fo}^2 + \left(\frac{\partial \bar{c}}{\partial \xi}\right)^2 \sigma_{\xi}^2 + \left(\frac{\partial \bar{c}}{\partial \xi_0}\right)^2 \sigma_{\xi_0}^2}. \tag{B.15}$$

Hence, the additional errors lower the local Fisher information: The stronger the sensitivities $(\partial\bar{c}/\partial Fo)$, $(\partial\bar{c}/\partial\xi)$, and $(\partial\bar{c}/\partial\xi_0)$ of the concentration $\bar{c}$ on the erroneous independent variables Fo, ξ, and ξ_0, and the larger the uncertainties σ_{Fo}, σ_ξ, and σ_{ξ_0} in the setting of the independent variables, the smaller the local Fisher information. Again, the local Fisher information is proportional to $(1/D)^2$. Therefore, the D-efficiencies are independent of the values of the diffusion coefficient D (cf. Equation 3.23).

For the subsequent example calculation, the following errors in the independent variables are assumed:

- $\sigma_{\bar{c}} = 0.02$.
- $\sigma_{Fo} = 0.005$. This corresponds to an error of 260 sec when $L = 12.5$ mm and $D = 3 \cdot 10^{-9}\,\mathrm{m}^2/\mathrm{sec}$.
- $\sigma_\xi = 0.04$. This corresponds to an error of 0.5 mm when $L = 12.5$ mm.
- $\sigma_{\xi_0} = 0.04$. This corresponds to an error of 0.5 mm when $L = 12.5$ mm.

For the example of the closed cell geometry with an initial domain boundary at position $\xi_0 = 0.5$, the contributions of the different error sources are shown in Figure B.2. The efficiencies are normalized to the case of a closed cell where no errors in the independent variables occur.

Errors in the Fourier number Fo (Figure B.2, left) lead to a loss of 11 % efficiency for the optimal conditions. More importantly, errors in the Fo-number mean in practice that the measurement time is not known with high accuracy. Thus, the likelihood to measure at the optimal time at the specified location is drastically reduced. For free diffusion, the range of measurement times leading to high efficiency is very small as can be seen from the narrow free diffusion ridge in Figure 5.2(a) (cf. Equation (5.13)). Hence, the free diffusion ridge at small *Fo*-numbers disappears if errors in the *Fo*-number are in taken into account. Thus, errors in the time measurement are particularly disadvantageous for free diffusion experiments.

Errors in the measurement position ξ (Figure B.2, center) lead to no loss of efficiency at the optimal measurement time and position because the optimal measurement position is already at the walls. However, errors in ξ mean that we do not know the measurement position precisely. Thus, the likelihood to measure at the optimal location for a specified time is again drastically reduced. This uncertainty has a large impact since the range of measurement locations leading to high efficiency is very small as can be seen from the narrow free diffusion ridge in Figure 5.2(a) (cf. Equation (5.13)). Errors in ξ therefore lead to the disappearance of the free diffusion ridge. Therefore, errors in ξ are also disadvantageous for free diffusion experiments.

Errors in the position ξ_0 of the initial domain boundary lead to high efficiency losses: 88 % of efficiency are lost for the optimal measurement time and position. The free diffusion ridge disappears again.

Hence, errors in the position ξ_0 of the initial domain boundary contribute most to efficiency losses, errors in the *Fo*-number the second most, and errors in the measurement position ξ the least. The most commonly employed free diffusion experiments are strongly affected by errors in the independent variables.

In general, the following characteristics are observed:

- For closed geometries, errors in the position ξ_0 of the initial domain boundary are the main cause for efficiency losses.
- For open geometries, errors in the Fourier number *Fo* are the main cause for efficiency losses because ξ_0 is no variable in these geometries.
- Both error contributions in ξ_0 and *Fo* increase from free/semi-free diffusion over cell and cylinder to the sphere. Thus, this decrease in efficiency due to errors in the independent variables runs in the opposite direction as the increase of efficiency due to walls and radial distortion (cf. Section 5.2.1). Hence, there is a trade-off between both effects; the specific result depends on the magnitude of errors in the independent variables.
- Errors in the measurement position ξ are of minor importance for optimal measurements, because the walls are always the optimal measurement position.
- The free diffusion ridge for small *Fo*-numbers disappears for all geometries.

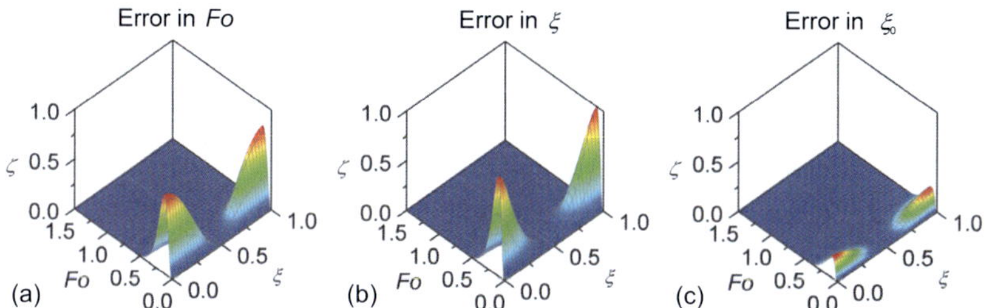

Figure B.2: Information in diffusion experiments as measured by the D-efficiencies ζ_D (Equation 3.23) of the closed cell with consideration of errors in the independent variables time Fo, measurement position ξ, and initial boundary position ξ_0 as a function of dimensionless position ξ, and dimensionless time (represented by the Fourier number Fo). The D-efficiencies are normalized to the case of a closed cell where no errors in the independent variables occur.

Hence, free diffusion experiments are particularly severely affected by errors in the independent variables.

The optimization results are shown in Figure B.3 for the case when errors in the independent variables are considered. Again, the optimization included the optimal position ξ_0 of the initial domain boundary. The efficiencies are normalized to the ideal case where no errors in the independent variables occur (cf. Section 5.2.2). Detailed values of the D-efficiencies are listed in Table B.4. The results show that open geometries are still the most beneficial geometries in general. The open cell shows the highest efficiencies in most cases. However, the efficiencies of the open sphere, which was the former optimal geometry, are severely affected by errors in the independent variables. Hence, the open cell is the most robust diffusion geometry.

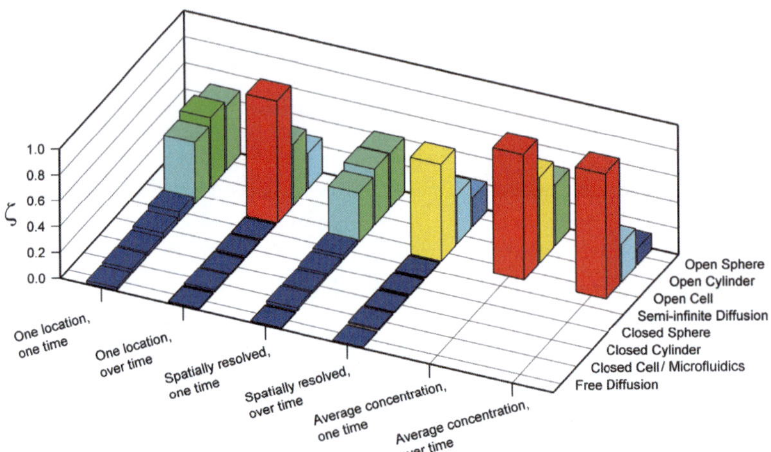

Figure B.3: Information in diffusion experiments as measured by the D-efficiencies ζ_{D} (Equation 3.23) for the optimal experiments with errors in the independent variables for each geometry and all spatial and temporal resolutions. Due to different objective functions, only results within one column can be compared to each other. The efficiencies are normalized to the case where no errors in the independent variables occur. Detailed values of the D-efficiencies are listed in Table B.4.

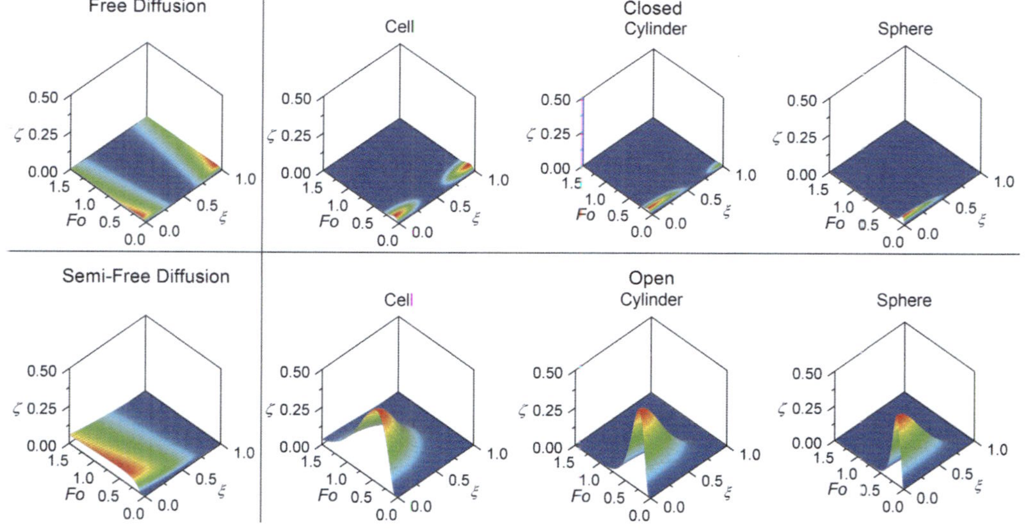

Figure B.4: D-efficiencies ζ_D (Equation 3.23) as a function of dimensionless position ξ and dimensionless time (represented by the Fourier number Fo) for the case of errors in the independent variables. The D-efficiencies are scaled to the maximum D-efficiency of the open sphere for the case when no errors in the independent variables occur. To allow for a clear representation, the positions ξ_0 of the initial domain boundary of the free diffusion geometry and the closed geometries (upper row) are not optimized here and chosen such that equal amounts of material are on both sides of the initial domain boundary.

Table B.4: Information in diffusion experiments as measured by the D-efficiencies ζ_D (Equation 3.23) for the optimal experiments with errors in the independent variables for each geometry and all spatial and temporal resolutions. Due to different objective functions, only results within one column can be compared to each other. The efficiencies are normalized to the case where no errors in the independent variables occur.

Geometry	One position, one time	One position, over time	Spatially resolved, one time	Spatially resolved, over time	Average concentration, one time	Average concentration, over time
Free diffusion	0.02*	0.01*	0.01*	0.00*	-	-
Closed cell/ Microfluidics	0.03	0.01	0.02	0.01	-	-
Closed cylinder	0.03	0.01	0.02	0.01	-	-
Closed sphere	0.04	0.01	0.02	0.01	-	-
Semi-free diffusion	0.07*	0.01*	0.03*	0.01*	-	-
Open cell	0.48	**0.94**	0.38	**0.75**	**0.97**	**0.96**
Open cylinder	**0.54**	0.47	**0.43**	0.38	0.66	0.29
Open sphere	0.49	0.27	0.42	0.23	0.47	0.12

*Free diffusion and semi-free diffusion equations are valid only as long as the walls have no or little influence on the diffusion process. The exact solution, considering the influence of the walls, is given by the equations of the closed cell and the open cell, respectively. In the optimization, a maximum error of 1% between concentrations in the free diffusion/semi-free diffusion geometry in comparison to the closed cell/open cell was allowed.

Appendix C

Diffusion in gases: Loschmidt cell: Additional data

Table C.1: Literature data

Reference	Temperature in K	Pressure in MPa	Mole fraction	D in $10^{-5}\,m^2/s$	Relative uncertainty in D
Carson and Dunlop (1972)	300	0.08918	0.0340	7.195	0.2 %
	300	0.08831	0.0735	7.278	0.2 %
	300	0.08917	0.1250	7.231	0.2 %
	300	0.09052	0.1313	7.115	0.2 %
	300	0.08654	0.2107	7.487	0.2 %
	300	0.08923	0.3520	7.330	0.2 %
	300	0.08923	0.3924	7.338	0.2 %
	300	0.08917	0.4393	7.351	0.2 %
	300	0.08908	0.4957	7.391	0.2 %
	300	0.08929	0.6003	7.384	0.2 %
	300	0.08869	0.6649	7.447	0.2 %
	300	0.08905	0.7280	7.429	0.2 %
	300	0.08461	0.8422	7.815	0.2 %
	300	0.08692	0.9011	7.622	0.2 %
	300	0.08915	0.9660	7.459	0.2 %
Winkelmann (2007)	293.1	0.101325	0.25	6.214	1, 5%
/Kestin et al. (1984)[1]	293.1	0.101324	0.5	6.295	1, 5 %
	293.1	0.101325	0.75	6.263	1, 5 %
Hogervorst (1971)	300	0.101325	300	6.32	$0.5 - 1, 5\,\%$
Staker and Dunlop (1976)	300	0.20265	0.025	0.3203	–
Staker et al. (1974)	300	0.06584099	0.051	0.9781	0.2 %
	300	0.06583085	0.1012	0.9827	0.2 %
	300	0.06582072	0.2024	0.9887	0.2 %
	300	0.06579032	0.4	0.9977	0.2 %
	300	0.06579032	0.5	1.00006	0.2 %
	300	0.06583085	0.7466	1.004	0.2 %
	300	0.06578019	0.8	1.0069	0.2 %
	300	0.06583085	0.8716	1.0074	0.2 %
	300	0.06580046	0.8945	1.0092	0.2 %
	300	0.06585112	0.9349	1.0092	0.2 %
	300	0.06579032	0.9465	1.0083	0.2 %
	300	0.06587138	0.967	1.0108	0.2 %
	300	0.06587138	0.9729	1.008	0.2 %
Srivastava and Paul (1962)	304.65	0.101325	0	6.6	1 %

[1] The data of Kestin et al. Kestin et al. (1984) is a correlation based on a large dataset from the

Table C.2: Parameters of the measurement setup and the system He-Kr

Parameter	Value	Reference
Length of the Loschmidt cell L in mm	397.14	Kugler et al. (2015b)
Depth of the Loschmidt cell l in mm	201	Kugler et al. (2015b)
Laser wavelength λ in nm	632.8	Kugler et al. (2015b)
Refractivity virial coefficient of Kr, $A_{R,Kr}$, in cm^3/mol	6.362	Achtermann et al. (1993)
Refractivity virial coefficient of He, $A_{R,He}$, in cm^3/mol	0.5213	Achtermann et al. (1993)

Table C.3: Diffusion coefficients D at 0.2 MPa and 293.15 K from the experiments with 30 mol% initial composition difference between the half-cells of the Loschmidt cell

Mole fraction of Kr in the lower half-cell	Mole fraction of Kr in the upper half cell	D with 95 % confidence interval in $10^{-5}\,m^2/s$
0.3	0	3.126 ± 0.006
0.6	0.3	3.142 ± 0.006
0.7	0.4	3.165 ± 0.008
1	0.7	3.182 ± 0.008

Table C.4: Estimated concentration-dependent diffusion coefficient $D = D^{(1)} \cdot x_1 + D^{(2)} \cdot (1 - x_1)$ of the system He(2)-Kr(1) at 0.2 MPa and 293.15 K from the individual fits of experiments (b1) and (b2)

Experiment	$D^{(1)}$ in $10^{-}5\,m^2/s$	$D^{(2)}$ in $10^{-}5\,m^2/s$
(b1)	3.0463	3.1667
(b2)	3.0573	3.2198
Mean value ±95 % confidence	3.117 ±0.051	3.069 ±0.079

principle of corresponding states, kinetic theory results, and experiments.

Appendix D

Diffusion in liquids: Microfluidic experiments: Additional data

D.1 Pure component data

Table D.1: Used chemicals with supplier and purity

	supplier	description	purity in %
cyclohexane	VWR	Spectronorm	99.7
toluene	VWR	Spectronorm	99.8
methanol	VWR	Spectronorm	99.9
acetone	Merck Millipore	Uvasol	99.9
1-propanol	Merck	LiChrosolv	99.8
1-chlorobutane	Merck	LiChrosolv	99.8
heptane	Bernd Kraft	p.a.	99

Table D.2: Molar volumes of pure components

	V_i^0 in $m^3\,mol^{-1}$	reference
cylohexane	108.8×10^{-6}	Beg et al. (1993)
toluene	106.9×10^{-6}	Chorążewski et al. (2010)
1-propanol	75.17×10^{-6}	Vercher et al. (2007)
1-chlorobutane	105.2×10^{-6}	Bolotnikov et al. (2007)
heptane	147.4×10^{-6}	Landaverde-Cortes et al. (2007)

D.2 Reparameterization for the estimation procedure of ternary diffusion coefficients

For ternary mixtures, the following constraints impose positive definiteness of the diffusion coefficient $\boldsymbol{D}^V$:Taylor and Krishna (1993)

$$D_{11} + D_{22} > 0 \tag{D.1}$$

$$D_{11}D_{22} - D_{12}D_{21} > 0 \tag{D.2}$$

$$(D_{11} - D_{22})^2 + 4D_{12}D_{21} > 0 \tag{D.3}$$

In practice, many solvers employ infeasible path strategies and allow violations of constraints during the iterations. We employed the reparameterization proposed by Bardow *et al.*:Bardow et al. (2006)

$$\theta_1 = D_1^\dagger,\ \theta_2 = D_2^\dagger,\ \theta_3 = D_{11} - D_{22},\ \text{and}\ \theta_4 = D_{12}\ . \tag{D.4}$$

The inverse transformation is performed according to

$$D_{11} = \frac{\theta_1 + \theta_2 + \theta_3}{2},\ D_{22} = \frac{\theta_1 + \theta_2 - \theta_3}{2},\ D_{21} = \frac{(\theta_1 - \theta_2)^2 - \theta_3^2}{4\theta_4},\ \text{and}\ D_{12} = \theta_4. \tag{D.5}$$

With simple constraints on θ_1 and θ_2, the constraints (D.1) - (D.3) are satisfied.Bardow et al. (2006) In practical applications, division by the unknown parameter θ_4 to compute D_{21} does not lead to numerical difficulties in our experience.

D.3 1-Propanol + 1-chlorobutane + heptane

Table D.3: Compositions of the experiments with 1-propanol + 1-chlorobutane + heptane in mol %

		series A		series B	
	$\bar{x}$	x_{left}	x_{right}	x_{left}	x_{right}
1-propanol	33.3	35.8	30.8	35.8	30.8
1-chlorobutane	33.3	35.8	30.8	30.8	35.8
heptane	33.3	28.4	38.4	33.4	33.4

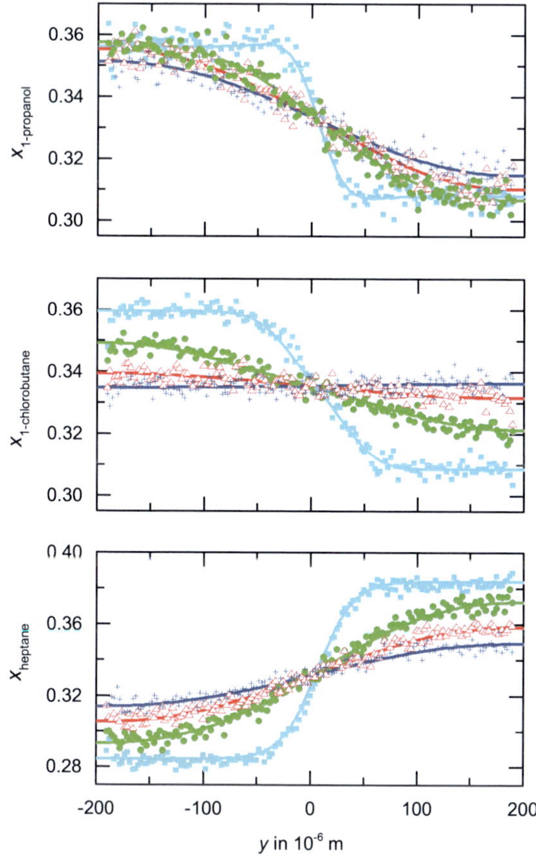

Figure D.1: Measured mole fraction profiles (symbols) and fitted mole fraction profiles (lines) of the system 1-propanol + 1-chlorobutane + heptane along measurement line y of the channel cross-sections at observation points $s_1 = 0.0143\,\text{m}$ ($\bar{t} = s/\bar{v} = 0.24\,\text{s}$, ■, –), $s_2 = 0.225\,\text{m}$ ($\bar{t} = 3.71\,\text{s}$, •, –), $s_3 = 0.452\,\text{m}$ ($\bar{t} = 7.44\,\text{s}$, △, –), and $s_4 = 0.694\,\text{m}$ ($\bar{t} = 11.43\,\text{s}$, +, –).

Table D.4: Measured diffusion coefficient for 1-propanol + 1-chlorobutane + heptane at $x_{\text{1-propanol}} = x_{\text{1-chlorobutane}} = 33.3\,\text{mol}\,\%$ and 25 °C by microfluidic experiment in this work with standard deviation $\sigma(\boldsymbol{D}^V)$ from repeated experiments in comparison to Käshammer et al. (1994).

	D_{11}	D_{12}	D_{21}	D_{22}
	in $1 \times 10^{-9}\,\text{m}^2\,\text{s}^{-1}$			
$n_{\text{exp}} = 1$[a]	0.746 ± 0.202	-0.552 ± 0.474	0.423 ± 0.120	2.571 ± 0.263
$n_{\text{exp}} = 2$[b]	0.998 ± 0.025	-0.279 ± 0.044	0.311 ± 0.027	2.521 ± 0.050
$n_{\text{exp}} = 9$[c]	1.017	-0.236	0.310	2.516
literature	1.033 ± 0.016	-0.174 ± 0.025	0.255 ± 0.012	2.463 ± 0.017

[a] estimated from nine separate experiments

[b] estimated from all combinations of two experiments, one from series A and series B each

[c] estimated from all experiments simultaneously, four from series A and five from series B each. No standard deviations from repeated experiments are available, but standard deviations are expected to be less than for $n_{\text{exp}} = 2$ as more data was used.

APPENDIX E

UNIQUAC and NRTL models and derivatives

E.1 NRTL

The NRTL model equations are Renon and Prausnitz (1968):

$$\ln\gamma_i = \frac{\sum_{j=1}^{n_c} \tau_{ji} G_{ji} x_j}{\sum_{l=1}^{n_c} G_{li} x_l} + \sum_{j=1}^{n_c} \frac{x_j G_{ij}}{\sum_{l=1}^{n_c} G_{lj} x_l} \left(\tau_{ij} - \frac{\sum_{n=1}^{n_c} x_n \tau_{nj} G_{nj}}{\sum_{l=1}^{n_c} G_{lj} x_l} \right) \tag{E.1}$$

where

$$\tau_{ij} = \frac{\tilde{g}_{ij}}{RT} \tag{E.2}$$

$$G_{ij} = \exp\left(-\alpha_{ij} \frac{\tilde{g}_{ij}}{RT} \right). \tag{E.3}$$

E.1.1 Derivatives of $\ln \gamma_i$ with respect to x_k and $\tilde{g}_{pq}$

The derivatives of the NRTL equations needed for the calculation of the Fisher matrix become:

$$\left.\frac{\partial \ln \gamma_i}{\partial x_k}\right|_{x_{r\neq k}} = \frac{\tau_{ki} G_{ki} \left[\sum_{l=1}^{n_c} G_{li} x_l\right] - \left[\sum_{j=1}^{n_c} \tau_{ji} G_{ji} x_j\right] G_{ki}}{\left[\sum_{l=1}^{n_c} G_{li} x_l\right]^2} + \sum_{j=1}^{n_c} \left[\frac{[\delta_{jk} G_{ik}] \left[\sum_{l=1}^{n_c} G_{lj} x_l\right] - x_j G_{ij} G_{kj}}{\left[\sum_{l=1}^{n_c} G_{lj} x_l\right]^2} \left(\tau_{ij} - \frac{\sum_{n=1}^{n_c} x_n \tau_{nj} G_{nj}}{\sum_{l=1}^{n_c} G_{lj} x_l} \right) - \frac{x_j G_{ij}}{\sum_{l=1}^{n_c} G_{lj} x_l} \frac{\tau_{kj} G_{kj} \left[\sum_{l=1}^{n_c} G_{lj} x_l\right] - \left[\sum_{n=1}^{n_c} x_n \tau_{nj} G_{nj}\right] G_{kj}}{\left[\sum_{l=1}^{n_c} G_{lj} x_l\right]^2} \right] \tag{E.4}$$

$$RT \frac{\partial \ln \gamma_i}{\partial \tilde{g}_{pq}} = \delta_{iq} \frac{(G_{pi} x_p - \tau_{pi} \alpha_{pq} G_{pq} x_p) \sum_{l=1}^{n_c} G_{li} x_l + \alpha_{pq} G_{pq} x_p \sum_{j=1}^{n_c} \tau_{ji} G_{ji} x_j}{\left[\sum_{l=1}^{n_c} G_{li} x_l\right]^2} + \frac{x_q G_{iq} \alpha_{pq} G_{pq} x_p - x_q \alpha_{pq} \delta_{ip} G_{pq} \sum_{l=1}^{n_c} G_{lq} x_l}{\left[\sum_{l=1}^{n_c} G_{lq} x_l\right]^2} \left(\tau_{iq} - \frac{\sum_{n=1}^{n_c} x_n \tau_{nq} G_{nq}}{\sum_{l=1}^{n_c} G_{lq} x_l} \right) + \frac{x_q G_{iq}}{\sum_{l=1}^{n_c} G_{lq} x_l} \left(\delta_{ip} - \frac{x_p G_{pq} (1 - \tau_{pq} \alpha_{pq}) \sum_{l=1}^{n_c} G_{lq} x_l + \alpha_{pq} G_{pq} x_p \sum_{n=1}^{n_c} x_n \tau_{nq} G_{nq}}{\left[\sum_{l=1}^{n_c} G_{lq} x_l\right]^2} \right) \tag{E.5}$$

E.1.2 UNIQUAC

The UNIQUAC model equations are Abrams and Prausnitz (1975):

$$\ln\gamma_i = \ln\gamma_i^{\mathrm{C}} + \ln\gamma_i^{\mathrm{R}} \tag{E.6}$$

where

$$\ln\gamma_i^{\mathrm{C}} = \ln\frac{\varphi_i}{x_i} + \frac{z}{2}q_i\ln\frac{\psi_i}{\varphi_i} + l_i - \frac{\varphi_i}{x_i}\sum_{j=1}^{n_c} x_j l_j \tag{E.7}$$

$$\ln\gamma_i^{\mathrm{R}} = q_i\left[1 - \ln\left(\sum_{j=1}^{n_c}\psi_j\tau_{ji}\right) - \sum_{j=1}^{n_c}\frac{\psi_j\tau_{ij}}{\sum_{l=1}^{n_c}\psi_l\tau_{lj}}\right] \tag{E.8}$$

with

$$l_i = \frac{z}{2}(r_i - q_i) - (r_i - 1) \tag{E.9}$$

$$\varphi_i = \frac{r_i x_i}{\sum_{j=1}^{n_c} r_j x_j} \tag{E.10}$$

$$\psi_i = \frac{q_i x_i}{\sum_{j=1}^{n_c} q_j x_j} \tag{E.11}$$

$$\tau_{ij} = \exp\left(-\frac{\tilde{u}_{ij}}{RT}\right). \tag{E.12}$$

E.1.3 Derivatives of $\ln\gamma_i$ with respect to x_k and $\tilde{\tau}_{pq}$

The derivatives of the UNIQUAC equations needed for the calculation of the Fisher matrix become:

$$\left.\frac{\partial\ln\gamma_i^{\mathrm{C}}}{\partial x_k}\right|_{x_{r\neq k}} = \frac{x_i}{\varphi_i}\frac{\frac{\partial\varphi_i}{\partial x_k}x_i - \varphi_i\delta_{ik}}{x_i^2} + \frac{z}{2}q_i\frac{\varphi_i}{\psi_i}\frac{\frac{\partial\psi_i}{\partial x_k}\varphi_i - \psi_i\frac{\partial\varphi_i}{\partial x_k}}{\varphi_i^2} - \frac{\frac{\partial\varphi_i}{\partial x_k}x_i - \varphi_i\delta_{ik}}{x_i^2}\left[\sum_{j=1}^{n_c}x_j l_j\right] - \frac{\varphi_i}{x_i}l_k \tag{E.13}$$

and

$$\left.\frac{\partial \ln \gamma_i^{\mathrm{R}}}{\partial x_k}\right|_{x_{r\neq k}} = q_i \left[-\frac{\sum_{j=1}^{n_c} \frac{\partial \psi_j}{\partial x_k} \tau_{ji}}{\sum_{j=1}^{n_c} \psi_j \tau_{ji}} - \sum_{j=1}^{n_c} \frac{\left[\frac{\partial \psi_j}{\partial x_k} \tau_{ij}\right] \left[\sum_{l=1}^{n_c} \psi_l \tau_{lj}\right] - \left[\psi_j \tau_{ij}\right] \left[\sum_{l=1}^{n_c} \frac{\partial \psi_l}{\partial x_k} \tau_{lj}\right]}{\left[\sum_{l=1}^{n_c} \psi_l \tau_{lj}\right]^2} \right] \tag{E.14}$$

where

$$\left.\frac{\partial \varphi_i}{\partial x_k}\right|_{x_{r\neq k}} = \frac{\left[r_i \delta_{ik}\right] \left[\sum_{j=1}^{n_c} r_j x_j\right] - \left[r_i x_i\right] r_k}{\left[\sum_{j=1}^{n_c} r_j x_j\right]^2} \tag{E.15}$$

and

$$\left.\frac{\partial \psi_i}{\partial x_k}\right|_{x_{r\neq k}} = \frac{\left[q_i \delta_{ik}\right] \left[\sum_{j=1}^{n_c} q_j x_j\right] - \left[q_i x_i\right] q_k}{\left[\sum_{j=1}^{n_c} q_j x_j\right]^2}. \tag{E.16}$$

Furthermore,

$$\frac{\partial \ln \gamma_i^{\mathrm{C}}}{\partial \tau_{pq}} = 0 \tag{E.17}$$

and

$$\frac{\partial \ln \gamma_i^{\mathrm{R}}}{\partial \tau_{pq}} = q_i \left[-\frac{\psi_p \delta_{iq}}{\sum_{j=1}^{n_c} \psi_j \tau_{ji}} - \frac{\psi_q \delta_{ip} \left[\sum_{l=1}^{n_c} \psi_l \tau_{lq}\right] - \psi_q \tau_{iq} \psi_p}{\left[\sum_{l=1}^{n_c} \psi_l \tau_{lq}\right]^2} \right] \tag{E.18}$$

Appendix F

OED of LLE: Additional data

F.1 LLE of type I with NRTL and LLE of type II with UNIQUAC and NRTL

In this section, the design of experiments for LLE of type I desribed by NRTL and for LLE of type II described by NRTL and UNIQUAC is presented in detail.

F.1.1 LLE of type I with NRTL

The continuous optimal design ξ^*_{opt} for the type I described by NRTL concentrates on 3 distinct positions:

$$\xi^*_{\text{opt}} = \left\{ \begin{array}{ccc} 0.34 & 0.63 & 0.8 \\ 0.25 & 0.37 & 0.38 \end{array} \right\}. \tag{F.1}$$

The exact optimal design $\xi^{(10)}_{\text{opt}}$ contains the following experiments:

$$\xi^{(10)}_{\text{opt}} = \left\{ \begin{array}{ccc} 0.34 & 0.63 & 0.8 \\ 0.2 & 0.4 & 0.4 \end{array} \right\}. \tag{F.2}$$

The conventional designs used as a benchmark for the optimal designs were the same as for the type I LLE described by UNIQUAC:

$$\xi^{(10)}_{\text{Conv1}} = \left\{ \begin{array}{ccc} 0 & 0.4 & 0.8 \\ 0.3 & 0.3 & 0.4 \end{array} \right\}, \tag{F.3}$$

$$\xi^{(10)}_{\text{Conv2}} = \left\{ \begin{array}{cccccccccc} 0 & 0.09 & 0.18 & 0.27 & 0.36 & 0.44 & 0.53 & 0.62 & 0.71 & 8 \\ 0.1 & 0.1 & 0.1 & 0.1 & 0.1 & 0.1 & 0.1 & 0.1 & 0.1 & 0.1 \end{array} \right\}, \tag{F.4}$$

Figure F.1 shows the predicted standardized variances d_{Pred} of the optimal designs ξ^*_{opt} and $\xi^{(10)}_{\mathrm{opt}}$ as well as of the conventional designs $\xi^{(10)}_{\mathrm{Conv1}}$ and $\xi^{(10)}_{\mathrm{Conv2}}$ plotted as a function of α. The values of the predicted standardized variance of the optimal design

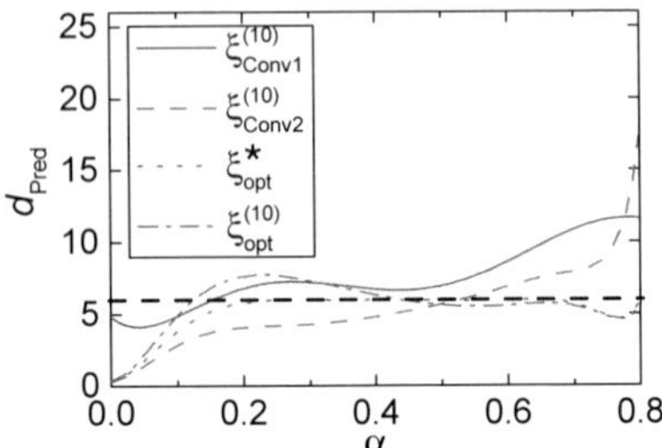

Figure F.1: Predicted standardized variances for the optimal and conventional experimental designs plotted against α. Designs based on NRTL.

ξ^*_{opt} do not exceed the value of 6, while the predicted standardized variance of the exact design $\xi^{(10)}_{\mathrm{opt}}$ barely exceeds 6. The values of the predicted standardized variance of the first conventional design $\xi^{(10)}_{\mathrm{Conv1}}$ are high where no experiment has been performed. The values of the predicted standardized variance of the second conventional design $\xi^{(10)}_{\mathrm{Conv2}}$ are low, except for the curved region near the critical point.

F.1.2 LLE of type II with UNIQUAC

The continuous optimal design ξ^*_{opt} for the type II LLE described by UNIQUAC concentrates on 5 distinct positions:

$$\xi^*_{\mathrm{opt}} = \left\{ \begin{array}{ccccc} 0 & 0.66 & 0.87 & 0.97 & 1 \\ 0.1276 & 0.2114 & 0.1934 & 0.1854 & 0.2822 \end{array} \right\}. \tag{F.5}$$

The exact optimal design $\xi^{(10)}_{\mathrm{opt}}$ contains the following experiments:

$$\xi^{(10)}_{\mathrm{opt}} = \left\{ \begin{array}{ccccc} 0 & 0.66 & 0.87 & 0.97 & 1 \\ 0.1 & 0.2 & 0.2 & 0.2 & 0.3 \end{array} \right\}. \tag{F.6}$$

As the optimal designs yielded 5 distinct experiments, the first conventional design also contains 5 distinct, although equally distributed, experiments:

$$\xi^{(10)}_{\text{Conv1}} = \left\{ \begin{array}{ccccc} 0 & 0.25 & 0.5 & 0.75 & 1 \\ 0.2 & 0.2 & 0.2 & 0.2 & 0.2 \end{array} \right\}, \tag{F.7}$$

$$\xi^{(10)}_{\text{Conv2}} = \left\{ \begin{array}{cccccccccc} 0 & 0.11 & 0.22 & 0.33 & 0.44 & 0.56 & 0.67 & 0.78 & 0.89 & 1 \\ 0.1 & 0.1 & 0.1 & 0.1 & 0.1 & 0.1 & 0.1 & 0.1 & 0.1 & 0.1 \end{array} \right\}. \tag{F.8}$$

Figure F.2 shows the positions of the equally distributed conventional experiments in the ternary phase diagrams.

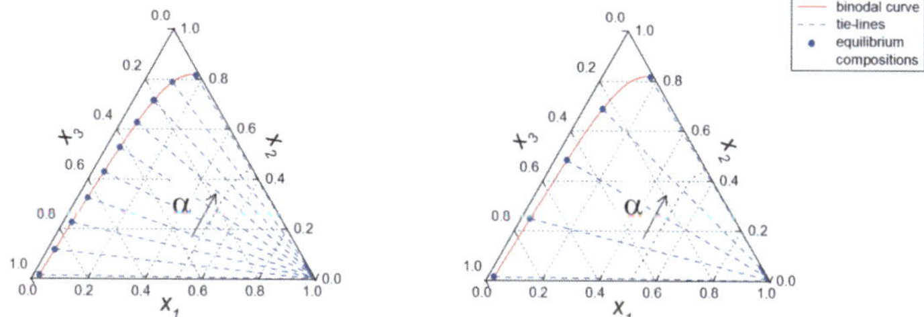

Figure F.2: Positions of the equally distributed conventional experiments in the ternary phase diagrams.
Left: $\alpha(\xi^{(10)}_{\text{Conv2}})$, using 10 distinct experiments; right: $\alpha(\xi^{(10)}_{\text{Conv1}})$, using 5 disctinct experiments. α is a linear scalar $[0, 1]$ to represent the overall compositions $\boldsymbol{z}$ in the middle of the tie-lines.

Figure F.3 shows the predicted standardized variances d_{Pred} of the optimal designs ξ^*_{opt} and $\xi^{(10)}_{\text{opt}}$ as well as of the conventional designs $\xi^{(10)}_{\text{Conv1}}$ and $\xi^{(10)}_{\text{Conv2}}$ plotted as a function of α.

We see that again, the values of the predicted standardized variance of the optimal designs ξ^*_{opt} do not exceed the value of 6. The values of predicted standardized variance barely exceed 6. However, the values of the predicted standardized variance of the first conventional design $\xi^{(10)}_{\text{Conv1}}$ are extremely high at regions where $\alpha > 0.7$. This is due to the absence of experiments where the curvature of the binodal curve is large. We observe the same effect for the values of the predicted standardized variance of the second conventional design $\xi^{(10)}_{\text{Conv2}}$, albeit in a less drastic way. For both conventional designs, the values of the predicted standardized variances for $\alpha < 0.7$ are lower than those for the optimal designs. This is due to the fact that below $\alpha = 0.7$, the course of

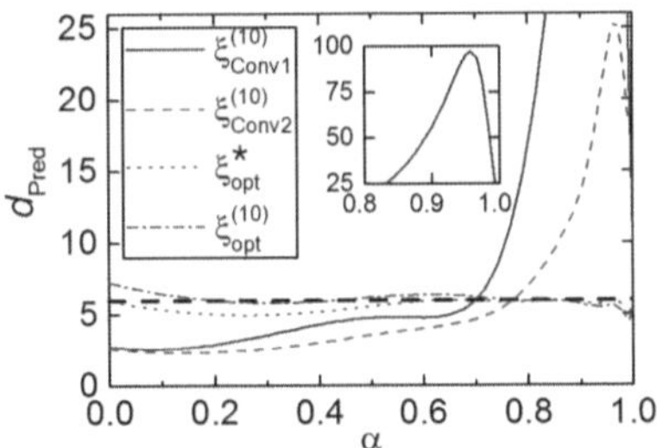

Figure F.3: Predicted standardized variances for the optimal and conventional experimental designs of type II LLE plotted against α. Designs based on UNIQUAC

the binodal curve is essentially linear. For the second conventional design $\xi^{(10)}_{\text{Conv2}}$, the low values of the predicted standardized variance result in comparable mean values to the mean values of the optimal designs.

F.1.3 LLE of type II with NRTL

The continuous optimal design ξ^{*}_{opt} for the type II LLE described by NRTL concentrates also on 5 distinct positions:

$$\xi^{*}_{\text{opt}} = \left\{ \begin{array}{ccccc} 0 & 0.61 & 0.82 & 0.93 & 1 \\ 0.1639 & 0.2429 & 0.1562 & 0.1291 & 0.3079 \end{array} \right\}. \tag{F.9}$$

The exact optimal design $\xi^{(10)}_{\text{opt}}$ contains the following experiments:

$$\xi^{(10)}_{\text{opt}} = \left\{ \begin{array}{ccccc} 0 & 0.61 & 0.82 & 0.93 & 1 \\ 0.2 & 0.2 & 0.2 & 0.1 & 0.3 \end{array} \right\}. \tag{F.10}$$

The conventional designs used as a benchmark for the optimal designs were the same as for the type II LLE described by UNIQUAC.

Figure F.4 shows the predicted standardized variances d_{Pred} of the optimal designs ξ^{*}_{opt} and $\xi^{(10)}_{\text{opt}}$ as well as of the conventional designs $\xi^{(10)}_{\text{Conv1}}$ and $\xi^{(10)}_{\text{Conv2}}$ plotted as a function of α. The values of the predicted standardized variances for the conventional designs are much lower than those for the type II LLE described by UNIQUAC.

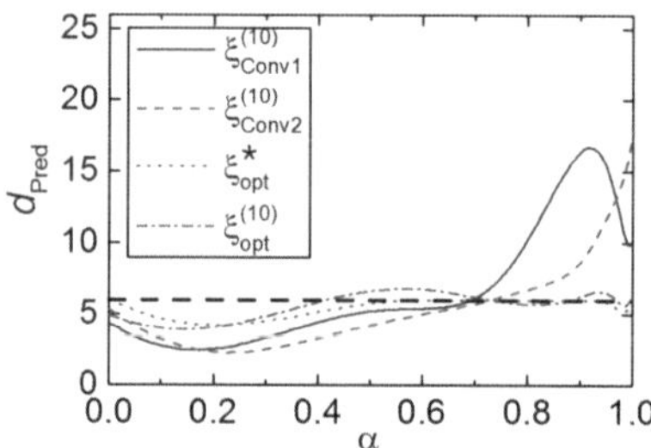

Figure F.4: Predicted standardized variances for the optimal and conventional experimental designs of type II LLE plotted against α. Designs based on NRTL

This even leads to lower mean values of the predicted standardized variances for the conventional designs than for the optimal designs (cf. Table 1 in the main document).

F.2 Model parameters

F.2.1 NRTL parameters

Type I: water(1)-acetone(2)-toluene(3) (Gmehling et al., 1979)

$$\tau = \begin{pmatrix} 0 & 0.7184 & 8.7235 \\ 1.2876 & 0 & -1.0285 \\ 4.4987 & 1.6688 & 0 \end{pmatrix}, \quad \alpha = const. = 0.2 \tag{F.11}$$

Type II: 3-methyltetrahydrofurane(1) - di-n-butylether(2) - water(3)

$$\tau = \begin{pmatrix} 0 & 2.6835 & 9.9947 \\ 0.6529 & 0 & 5.6044 \\ 2.6252 & -3.1213 & 0 \end{pmatrix}, \quad \alpha = const. = 0.2 \tag{F.12}$$

F.2.2 UNIQUAC parameters

Type I: water(1)-acetone(2)-toluene(3) (Gmehling et al., 1979)

$$\tau = \begin{pmatrix} 1 & 1.3400 & 0.5550 \\ 0.2600 & 1 & 1.6100 \\ 0.0344 & 0.4000 & 1 \end{pmatrix} \tag{F.13}$$

$$q = \begin{pmatrix} 1.400 & 2.336 & 2.968 \end{pmatrix}, \quad r = \begin{pmatrix} 0.9200 & 2.5735 & 3.9228 \end{pmatrix}, \quad z = 10 \tag{F.14}$$

Type II: 3-methyltetrahydrofurane(1) - di-n-butylether(2) - water(3)

$$\tau = \begin{pmatrix} 1 & 0.9521 & 0.4022 \\ 0.2621 & 1 & 0.2025 \\ 0.1184 & 2.3711 & 1 \end{pmatrix} \tag{F.15}$$

$$q = \begin{pmatrix} 1.40000 & 2.54955 & 5.17600 \end{pmatrix}, \quad r = \begin{pmatrix} 0.92000 & 2.92171 & 6.09250 \end{pmatrix}, \quad z = 10 \tag{F.16}$$

Type I with small miscibility gap:

$$\tau = \begin{pmatrix} 1 & 0.30 & 0.12 \\ 2.29 & 1 & 0.40 \\ 0.97 & 1.06 & 1 \end{pmatrix} \tag{F.17}$$

$$q = \begin{pmatrix} 3.240 & 2.968 & 1.432 \end{pmatrix}, \quad r = \begin{pmatrix} 4.0464 & 3.9228 & 1.4311 \end{pmatrix}, \quad z = 10 \tag{F.18}$$

Appendix G

OED for extraction processes: Additional data

G.1 Parameters

Table G.1: Process parameters

Parameter	Value	Reference
Mass flow of raffinate carrier water $\dot{m}_1''$ in kg/s	5	
Specific mass transfer area of packing a_P in m^2/m^3	150.92	Strigle (1994)
Void fraction area of packing ϵ	0.972	Strigle (1994)
Boundary layer thickness δ' in m	10^{-4}	
Diffusion coefficient of acetone in water D_{12} in m^2/s	$0.81 \cdot 10^{-9}$	Tyn and Calus (1975)
Diffusion coefficient of acetone in toluene D_{23} in m^2/s	$1.1 \cdot 10^{-9}$	Bulicka and Prochazka (1976)
Raffinate inlet loading $X''_{2,\mathrm{in}}$ in mol acetone/mol water	0.2	
Raffinate purification yield $X''_{2,\mathrm{out}}$ in mol acetone/mol water	0.05	
Extract inlet loading $X'_{2,\mathrm{in}}$ in mol acetone/mol water	0	
Molar mass water in kg/mol	0.01802	
Molar mass acetone in kg/mol	0.05808	
Molar mass toluene in kg/mol	0.09214	
Density of raffinate phase ρ'' in kg/m^3	997.05	NIST (2018)
Density of extract phase ρ' in kg/m^3	862.24	NIST (2018)
Viscosity of raffinate phase μ'' in Pa s	0.00089	NIST (2018)
Viscosity of extract phase μ' in Pa s	0.00056	NIST (2018)
Interfacial tension σ_I in N/m	$25.46 \cdot 10^{-7}$	Enders et al. (2007)
Safety factor $\mathcal{S}$ for Crawford-Wilke correlation	5	Strigle (1994)

Table G.2: Cost parameters

Parameter	Value	Reference
Plant lifetime N_{life} in a	25	
Annual operation hours t_{a} in h	8000	
Interest rate I	4.18%	
Disposal costs C_{disp} in \$/kg	1	
Material factor F_{m}	2.25	Biegler et al. (1997)
Pressure factor F_{p}	1	Biegler et al. (1997)
Base costs F_{p}	1	Biegler et al. (1997)
Pressure factor F_{p}	1	Biegler et al. (1997)
Pressure factor F_{p}	1	Biegler et al. (1997)
Pressure factor F_{p}	1	Biegler et al. (1997)
Reference height of column $h_{\mathrm{K},0}$ in ft	4	Biegler et al. (1997)
Reference diameter of column $d_{\mathrm{K},0}$ in ft	3	Biegler et al. (1997)
Exponent for column height scaling α_{h}	0.81	Biegler et al. (1997)
Exponent for column diameter scaling α_{d}	1.05	Biegler et al. (1997)

Table G.3: Module factor MF as function of base costs BC (Biegler et al., 1997)

BC in 10^3 \$	0...20	20...400	400...600	600...800	800...1000
MF	4.23	4.12	4.07	4.06	4.02

G.2 Results when LLE are modeled with NRTL

$$\boldsymbol{d}^{*}_{\mathrm{c,opt}} = \left\{ \begin{array}{ccc|c|c} & \text{LLE} & & \text{Diffusion}' & \text{Diffusion}'' \\ \hline \alpha = 0.16 & \alpha = 0.61 & \alpha = 0.8 & \begin{array}{c} Fo = 0.1 \\ \xi = 0 \end{array} & \begin{array}{c} Fo = 0.1 \\ \xi = 0 \end{array} \\ \hline v = 0.82 & v = 0.04 & v = 0.02 & v = 0.09 & v = 0.03 \end{array} \right\}. \tag{G.1}$$

$$\boldsymbol{d}^{(12)}_{\mathrm{c,opt}} = \left\{ \begin{array}{ccc|c|c} & \text{LLE} & & \text{Diffusion}' & \text{Diffusion}'' \\ \hline \alpha = 0.16 & \alpha = 0.61 & \alpha = 0.06 & \begin{array}{c} Fo = 0.1 \\ \xi = 0 \end{array} & \begin{array}{c} Fo = 0.1 \\ \xi = 0 \end{array} \\ \hline v = 8/12 & v = 1/12 & v = 1/12 & v = 1/12 & v = 1/12 \end{array} \right\}. \tag{G.2}$$

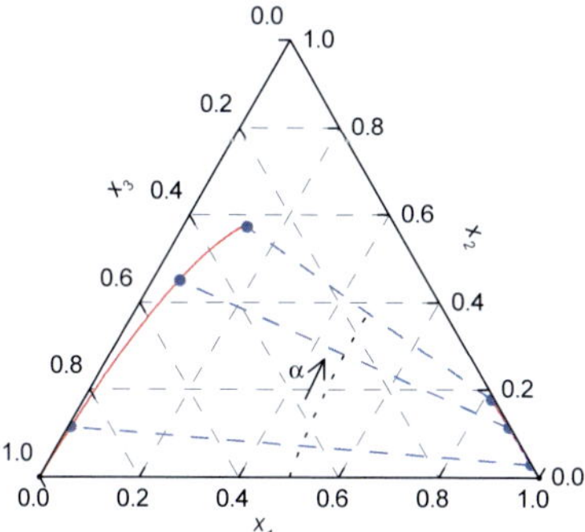

Figure G.1: Positions of the c-optimal LLE experiments for the rate-based extraction process in the ternary phase diagrams. The LLE is based on NRTL.

Table G.4: c-efficiencies ζ_c (Equation 3.23) of the continuous and exact c-optimal designs $\boldsymbol{d}^*$ and $\boldsymbol{d}^{(12)}_{\text{opt}}$, as well as of the D-optimal $\boldsymbol{d}^{(12)}_{\text{opt,LLE}}$ design for the exclusive determination of LLE parameters, and the conventional designs $\boldsymbol{d}^{(12)}_{\text{Conv1}}$ and $\boldsymbol{d}^{(12)}_{\text{Conv2}}$. The LLE is based on NRTL

LLE design	$\boldsymbol{d}^*$ (Eq. 8.29)	$\boldsymbol{d}^{(12)}_{\text{opt}}$ (Eq. 8.30)	$\boldsymbol{d}^{(12)}_{\text{opt,LLE}}$ (Eq. 7.3)	$\boldsymbol{d}^{(12)}_{\text{Conv1}}$ (Eq. 7.4)	$\boldsymbol{d}^{(12)}_{\text{Conv2}}$ (Eq. 7.5)
ζ_c	1	0.84	0.19	0.06	0.36

Bibliography

Abrams, D. S. and Prausnitz, J. M. (1975). Statistical thermodynamics of liquid mixtures: a new expression for the excess gibbs energy of partly or completely miscible systems. *AIChE J.*, 21(1):116–128.

Achtermann, H. J., Hong, J. G., Magnus, G., Aziz, R. A., and Slaman, M. J. (1993). Experimental determination of the refractivity virial coefficients of atomic gases. *J. Chem. Phys.*, 98(3):2308–2318.

Albright, J. G. and Miller, D. G. (1975). Analysis of free diffusion in a binary system when the diffusion coefficient is a function of the square root of concentration. *J. Phys. Chem.*, 79:2061–2068.

Albright, J. G. and Miller, D. G. (1980). Analysis of Gouy interference patterns from binary free-diffusion systems when the diffusion coefficient and refractive index have c1/2 and c3/2 terms, respectively. *J. Phys. Chem.*, 84:1400–1413.

Alimuddin, M., Grant, D., Bulloch, D., Lee, N., Peacock, M., and Dahl, R. (2008). Determination of log D via automated microfluidic liquid–liquid extraction. *J. Med. Chem.*, 51(16):5140–5142.

Allen, M. P. and Tildesley, D. J. (2017). *Computer Simulation of Liquids.* Oxford University Press, Croydon, 2nd edition.

Allie-Ebrahim, T., Russo, V., Ortona, O., Paduano, L., Tesser, R., Di Serio, M., Singh, P., Zhu, Q., Moggridge, G., and D'Agostino, C. (2018). A predictive model for the diffusion of a highly non-ideal ternary system. *Phys. Chem. Chem. Phys.*, 20:18436–18446.

Alsmeyer, F., Koß, H.-J., and Marquardt, W. (2004). Indirect spectral hard modeling for the analysis of reactive and interacting mixtures. *Appl. Spectrosc.*, 58(8):975–985.

Amalberti, J., Antoine, X., and Burnard, P. (2018a). Timescale monitoring of vesuvian eruptions using numerical modeling of the diffusion equation. *Math. Geosci.*, 50(4):417–429.

Amalberti, J., Burnard, P., Tissandier, L., and Laporte, D. (2018b). The diffusion coefficients of noble gases (hear) in a synthetic basaltic liquid: One-dimensional diffusion experiments. *Chem. Geol.*, 480:35–43.

Ambrosone, L., D'Errico, G., Sartorio, R., and Vitagliano, V. (1995). Analysis of velocity cross-correlation and preferential solvation for the system n-methylpyrrolidone–water at 20°C. *J. Chem. Soc. Faraday T.*, 91(9):1339–1344.

Anderson, D. K. and Babb, A. L. (1961). Mutual diffusion in non-ideal liquid mixtures. II. diethyl ether–chloroform. *J. Phys. Chem.*, 65(8):1281–1283.

Anderson, D. K., Hall, J. R., and Babb, A. L. (1958). Mutual diffusion in non-ideal binary liquid mixtures. *J. Phys. Chem.*, 62:404–408.

Anderson, J. S. and Saddington, K. (1949). S 80. The use of radioactive isotopes in the study of the diffusion of ions in solution. *J. Chem. Soc.*, pages S381–S386.

Aoyagi, K. and Albright, J. G. (1972). Tracer diffusion and viscosity study at 25.deg. in binary and ternary liquid systems. *J. Phys. chem.*, 76(18):2572–2577.

Arora, P. S., Carson, P. J., and Dunlop, P. J. (1978). Determination of potential parameters for the systems Ne-Ar and Ar-Kr from the temperature dependence of their binary diffusion coefficients. *Chem. Phys. Lett.*, 54(1):117–119.

Atkinson, A. C., Donev, A. N., and Tobias, R. D. (2007). *Optimum experimental designs, with SAS.* Oxford University Press, Oxford.

Bacon, J. and Adams, R. N. (1970). Capillary diffusion measurements using fluorescence analysis. D values of some electrochemically important systems. *Anal. Chem.*, 42(4):524–525.

Balaji, S. P., Schnell, S. K., McGarrity, E. S., and Vlugt, T. J. H. (2013). A direct method for calculating thermodynamic factors for liquid mixtures using the permuted widom test particle insertion method. *Mol. Phys.*, 111(2):287–296.

Baranski, J. (2002). *Bestimmung binärer Diffusionskoeffizienten von Gasen mit einer Loschmidt-Zelle und holografischer Interferometrie.* Dissertation, Universität Rostock, Rostock.

Bard, Y. (1974). *Nonlinear parameter estimation.* Academic Press.

Bardow, A. (2004). *Model-based Experimental Analysis of Multicomponent Diffusion.* Dissertation, RWTH Aachen University.

Bardow, A. (2007). On the interpretation of ternary diffusion measurements in low-molecular weight fluids by dynamic light scattering. *Fluid Phase Equilibr.*, 251:121–127.

Bardow, A., Göke, V., Koß, H.-J., Lucas, K., and Marquardt, W. (2005a). Concentration-dependent diffusion coefficients from a single experiment using model-based Raman spectroscopy. *Fluid Phase Equilibr.*, 228:357–366.

Bardow, A., Göke, V., Koß, H.-J., Lucas, K., and Marquardt, W. (2005b). Concentration-dependent diffusion coefficients from a single experiment using model-based raman spectroscopy. *Fluid Phase Equilibr.*, 228-229:357–366.

Bardow, A., Göke, V., Koß, H.-J., and Marquardt, W. (2006). Ternary diffusivities by model-based analysis of Raman spectroscopy measurements. *AIChE J.*, 52:4004–4015.

Bardow, A., Marquardt, W., Göke, V., Koß, H.-J., and Lucas, K. (2003). Model-based measurement of diffusion using Raman spectroscopy. *AIChE J.*, 49:323–334.

Barz, T., Arellano-Garcia, H., and Wozny, G. (2010). Handling uncertainty in model-based optimal experimental design. *Ind. Eng. Chem. Res.*, 49(12):5702–5713.

Bausa, J. and Marquardt, W. (2001). Detailed modeling of stationary and transient mass transfer across pervaporation membranes. *AIChE J.*, 47(6):1318–1332.

Beg, S. A., Tukur, N. M., Al-Harbi, D. K., and Hamad, E. Z. (1993). Saturated liquid densities of benzene, cyclohexane, and hexane from 298.15 to 473.15 K. *J. Chem. Eng. Data*, 38(3):461–464.

Ben-Naim, A. (2006). *Molecular Theory of Solutions.* Oxford University Press, Oxford.

Berg, R. W., Brunsgaard Hansen, S., Shapiro, A. A., and Stenby, E. H. (2007). Diffusion measurements in binary liquid mixtures by Raman spectroscopy. *Appl. Spectrosc.*, 61(4):367–373.

Bernardo, F., Saraiva, P., and Pistikopoulos, E. (2003). Process design under uncertainty: robustness criteria and value of information. *Comput.-Aided Chem. En.*, 16:175–208.

Bernardo, F. P., Pistikopoulos, E. N., and Saraiva, P. M. (2001). Quality costs and robustness criteria in chemical process design optimization. *Comput. Chem. Eng.*, 25(1):27–40.

Bernardo, F. P., Saraiva, P., and Efstratios, P. N. (2000). Inclusion of information costs in process design optimization under uncertainty. *Comput. Chem. Eng.*, 24(2-7):1695–1701.

Bernardo, F. P. and Saraiva, P. M. (2004). Value of information analysis in product/process design. *Comput.-Aided Chem. En.*, 18:151–156.

Bernardo, F. P. and Saraiva, P. M. (2015). A conceptual model for chemical product design. *AIChE J.*, 61(3):802–815.

Bidlack, D. and Anderson, D. (1964). Mutual diffusion in nonideal, nonassociating liquid systems. *J. Phys. Chem.*, 68:3790–3794.

Biegler, L., Grossmann, I., and Westerberg, A. (1997). *Systematic Methods for Chemical Process Design.* Prentice Hall, Old Tappan, New Jersey (United States).

Bird, R. B. (2004). Five decades of transport phenomena. *AIChE J.*, 50(2):273–287.

Bird, R. B., Stewart, W. E., and Lightfoot, E. N. (2002). *Transport Phenomena.* John Wiley & Sons, 2nd edition.

Bolden, R. D., Hoke II, S. H., Eichhold, T. H., McCauley-Myers, D. L., and Wehmeyer, K. R. (2002). Semi-automated liquid–liquid back-extraction in a 96-well format to decrease sample preparation time for the determination of dextromethorphan and dextrorphan in human plasma. *J. Chromatogr. B*, 772(1):1–10.

Bolotnikov, M. F., Neruchev, Y. A., and Ryshkova, O. S. (2007). Density of some 1-chloroalkanes within the temperature range from (253.15 to 423.15) k. *J. Chem. Eng. Data*, 52(6):2514–2516.

Bosse, D. and Bart, H.-J. (2005). Measurement of diffusion coefficients in thermodynamically nonideal systems. *J. Chem. Eng. Data*, 50(5):1525–1528.

Bouchaudy, A., Loussert, C., and Salmon, J.-B. (2018). Steady microfluidic measurements of mutual diffusion coefficients of liquid binary mixtures. *AIChE J.*, 64(1):358–366.

Boyd, C. A., Stein, N., Steingrimsson, V., and Rumpel, W. F. (1951). An interferometric method of determining diffusion coefficients in gaseous systems. *J. Chem. Phys.*, 19(5):548–553.

Bulicka, J. and Prochazka, J. (1976). Diffusion coefficients in some ternary systems. *J. Chem. Eng. Data*, 21(4):452–456.

Buttig, D. (2010). *Bestimmung binärer Diffusionskoeffizienten in Gasmischungen mit einer Loschmidt-Zelle und holografischer Interferometrie*. Dissertation, Universität Rostock.

Buttig, D., Vogel, E., Bich, E., and Hassel, E. (2011). A Loschmidt cell combined with holographic interferometry for binary diffusion experiments in gas mixtures including first measurements on the argon–neon system. *Meas. Sci. Technol.*, 22(10):105409.

Caldwell, C. S. and Babb, A. L. (1956). Diffusion in ideal binary liquid mixtures. *J. Phys. Chem.*, 60:51–56.

Carlsson, K. and Karlberg, B. (2000). Determination of octanol–water partition coefficients using a micro-volume liquid-liquid flow extraction system. *Anal. Chim. Acta*, 423(1):137–144.

Carman, P. C. and Stein, L. H. (1956). Self-diffusion in mixtures. Part 1. Theory and its application to a nearly ideal binary liquid mixture. *Trans. Faraday Soc.*, 52:619–627.

Carslaw, H. S. and Jaeger, J. C. (1959). *Conduction of Heat in Solids*. Clarendon Press, Oxford, 2nd edition.

Carson, P. J. and Dunlop, P. J. (1972). The composition dependence of the diffusion coefficient of the system helium-krypton at 300K and 1 atmosphere pressure. *Chem. Phys. Lett.*, 14(3):377–379.

Case, F. H., Chaka, A., Moore, J. D., Mountain, R. D., Ross, R. B., Shen, V. K., and Stahlberg, E. A. (2011). The sixth industrial fluid properties simulation challenge. *Fluid Phase Equilibr.*, 310(1-2):1–3.

Castañer, J. and Ramírez, C. A. (2018). Binary gas diffusivity estimates from transient, one-dimensional sublimation-diffusion experiments in a spherical enclosure. *Chem. Eng. Commun.*, 205(9):1167–1192.

Chang, L.-C., Lin, T.-I., and Li, M.-H. (2005). Mutual diffusion coefficients of some aqueous alkanolamines solutions. *J. Chem. Eng. Data*, 50:77–84.

Chorążewski, M., Grolier, J.-P. E., and Randzio, S. L. (2010). Isobaric thermal expansivities of toluene measured by scanning transitiometry at temperatures from (243 to 423) K and pressures up to 200 MPa. *J. Chem. Eng. Data*, 55(12):5489–5496.

Chu, Y. and Hahn, J. (2008). Integrating parameter selection with experimental design under uncertainty for nonlinear dynamic systems. *AIChE J.*, 54(9):2310–2320.

Clark, W. M. and Rowley, R. L. (1988). The mutual diffusion coefficient of methanol–n-hexane near the consolute point. *AIChE J.*, 32(7):1125–1131.

Clunie, J. C., Li, N., Emerson, M. T., and Baird, J. K. (1990). Theory and measurement of the concentration dependence of the differential diffusion coefficient using a diaphragm cell with compartments of unequal volume. *J. Phys. Chem.*, 94(15):6099–6105.

Costin, C. D., Olund, R. K., Staggemeier, B. A., Torgerson, A. K., and Synovec, R. E. (2013). Diffusion coefficient measurement in a microfluidic analyzer using dual-beam microscale-refractive index gradient detection: Application to on-chip molecular size determination. *J. Chromatogr. A*, 1013:77–91.

Crank, J. (1975). *The mathematics of diffusion.* Clarendon press, Oxford.

Cullinan, H. T. and Toor, H. L. (1965). Diffusion in the three-component liquid system acetone–benzene–carbon tetrachloride. *J. Phys. Chem.*, 69:3941–3949.

Curtiss, C. F. and Hirschfelder, J. O. (1949). Transport properties of multicomponent gas mixtures. *J. Chem. Phys*, 17(6):550–555.

Cussler, E. L. (2007). *Diffusion - Mass Transfer in Fluid Systems.* Cambridge.

D'Agostino, C., Mantle, M. D., Gladden, L., and Moggridge, G. (2011). Prediction of binary diffusion coefficients in non-ideal mixtures from NMR data: Hexane–nitrobenzene near its consolute point. *Chem. Eng. Sci.*, 66:3898–3906.

D'Agostino, C., Mantle, M. D., Gladden, L. F., and Moggridge, G. D. (2012). Prediction of mutual diffusion coefficients in non-ideal mixtures from pulsed field gradient NMR data : Triethylamine–water near its consolute point. *Chem. Eng. Sci.*, 74:105–113.

D'Agostino, C., Stephens, J. A., Parkinson, J. D., Mantle, M. D.and Gladden, L. F., and Moggridge, G. D. (2013). Prediction of the mutual diffusivity in acetone-chloroform liquid mixtures from the tracer diffusion coefficients. *Chem. Eng. Sci.*, 95:43–47.

Darken, L. S. (1948). Diffusion, mobility and their interrelation through free energy in binary metallic systems. *Trans. Aime*, 175:184–201.

Dawass, N., Krüger, P., Schnell, S. K., Bedeaux, D., Kjelstrup, S., Simon, J. M., and Vlugt, T. J. H. (2017). Finite-size effects of Kirkwood-Buff integrals from molecular simulations. *Mol. Simul.*, 7022:1–14.

De, S., Shapir, Y., and Chimowitz, E. (2001). Scaling of self and fickian diffusion coefficients in the critical region. *Chem. Eng. Sci.*, 56(17):5003–5010.

Dechambre, D., Pauls, C., Greiner, L., Leonhard, K., and Bardow, A. (2014a). Towards automated characterisation of liquid–liquid equilibria. *Fluid Phase Equilibr.*, 362:328–334.

Dechambre, D., Wolff, L., Pauls, C., and Bardow, A. (2014b). Optimal experimental design for the characterization of liquid-liquid equilibria. *Ind. Eng. Chem. Res.*, 53(50):19620–19627.

Derlacki, Z. J., Easteal, A. J., Edge, A. V. J., and Woolf, L. A. (1985). Diffusion coefficients of methanol and water and the mutual diffusion coefficient in methanol–water solutions at 278 and 298 K. *J. Phys. Chem.*, 89(24):5318–5322.

DortmundDataBank (2019). http://ddbonline.ddbst.com, visited on 21 september 2019.

Durou, C., Moutou, C., and Mahenc, J. (1974). Détermination des coefficients de diffusion différentiels isothermes par interférométrie holographique et simulation numérique. *J. Chim. Phys.*, 71:271–277.

Easteal, A., Woolf, L., and Mills, R. (1987). Velocity cross-correlation coefficients for the system acetonitrile-water at 278 K and 298 K. *Z. Phys. Chem.*, 155(1):69–78.

Economou, I. G., de Hemptinne, J.-C., Dohrn, R., Hendriks, E., Keskinen, K., and Baudouin, O. (2014). Industrial use of thermodynamics workshop: Round table discussion on 8 July 2014. *Chem. Eng. Res. Des.*, 92:2795–2796.

Enders, S., Kahl, H., and Winkelmann, J. (2007). Surface tension of the ternary system water + acetone + toluene. *J. Chem. Eng. Data*, 52(3):1072–1079.

Estévez-Torres, A., Le Saux, T., Gosse, C., Lemarchand, A., Bourdoncle, A., and Jullien, L. (2008). Fourier transform to analyse reaction-diffusion dynamics in a microsystem. *Lab Chip*, 8:1205–1209.

Evans, D. J. and Morriss, G. P. (2008). *Statistical Mechanics of Nonequilibrium Liquids.* Cambridge University Press.

Fick, A. (1855). Über Diffusion. *Ann. Phys.*, 170:59–86.

Fredenslund, A., Jones, R. L., and Prausnitz, J. M. (1975). Group-contribution estimation of activity coefficients in nonideal liquid mixtures. *AIChE J.*, 21(6):1086–1099.

French, H. (1987). Vapour pressures and activity coefficients of (acetonitrile + water) at 308.15 K. *J. Chem. Thermodyn.*, 19(11):1155–1161.

Frenkel, D. and Smit, B. (2002). *Understanding Molecular Simulation: From Algorithms to Applications.* Academic Press, London, 2nd edition.

Friedman, H. and Mills, R. (1981). Velocity cross correlations in binary mixtures of simple fluids. *J. Solution Chem.*, 10(6):395–409.

Friedman, H. L. and Mills, R. (1986). Hydrodynamic approximation for distinct diffusion coefficients. *J. Solution Chem.*, 15(1):69–80.

Fujita, H. (1956). The zero-time correction in free diffusion experiments. *J. Phys. Soc. Jpn.*, 11:1018–1019.

Galvanin, F., Boschiero, A., Barolo, M., and Bezzo, F. (2011). Model-based design of experiments in the presence of continuous measurement systems. *Ind. Eng. Chem. Res.*, 50:2167–2175.

Ganguly, P. and van der Vegt, N. F. A. (2013). Convergence of sampling Kirkwood-Buff integrals of aqueous solutions with molecular dynamics simulations. *J. Chem. Theory Comput.*, 9(3):1347–1355.

Geiger, A., Hertz, H., and Mills, R. (1981). Velocity correlations in aqueous electrolyte solutions from diffusion, conductance and transference data: Application to concentrated solutions of nickel chloride and magnesium chloride. *J. Solution Chem.*, 10(2):83–94.

Ghai, R. K. and Dullien, F. A. L. (1974). Diffusivities and viscosities of some binary liquid nonelectrolytes at 25°. *J. Phys. Chem.*, 78:2283–2291.

Gmehling, J., Kolbe, B., Kleiber, M., and Rarey, J. (2012). *Chemical Thermodynamics for Process Simulation.* Wiley-VCR Verlag & Co. KGaA.

Gmehling, J., Onken, U., and Arlt, W. (1979). *Vapor-liquid equilibrium data collection.* Dechema, Frankfurt am Main.

Gordon, A. R. (1945). The diaphragm cell method of measuring diffusion. *Ann. NY Acad. Sci.*, 46(5):285–308.

Gotoh, S., Manner, M., Sørensen, J. P., and Stewart, W. E. (1973). Studies of the Loschmidt diffusion experiment. II. an improved interferometric method. *Ind. Eng. Chem. Fundament.*, 12(1):119–123.

Graham, T. (1850). The Bakerian lecture: On the diffusion of liquids. *Philos. Tr. Soc. Lond.*, 140:1–46.

Grossmann, T. and Winkelmann, J. (2005). Ternary diffusion coefficients of glycerol + acetone + water by Taylor dispersion measurements at 298.15 K. *J. Chem. Eng. Data*, 50:1396–1403.

Guevara-Carrion, G., Gaponenko, Y., Janzen, T., Vrabec, J., and Shevtsova, V. (2016a). Diffusion in multicomponent liquids: From microscopic to macroscopic scales. *J. Phys. Chem. B*, 120(47):12193–12210.

Guevara-Carrion, G., Janzen, T., Muñoz-Muñoz, Y. M., and Vrabec, J. (2016b). Mutual diffusion of binary liquid mixtures containing methanol, ethanol, acetone, benzene, cyclohexane, toluene, and carbon tetrachloride. *J. Chem. Phys.*, 144(12):124501.

Guevara-Carrion, G., Nieto-Draghi, C., Vrabec, J., and Hasse, H. (2008). Prediction of transport properties by molecular simulation: Methanol and ethanol and their mixture. *J. Phys. Chem. B*, 112:16664–16674.

Gupta, P. K. and Cooper, A. R. (1971). The [D] matrix for multicomponent diffusion. *Physica*, 54(1):39–59.

Haase, R. and Siry, M. (1968). Diffusion im kritischen Entmischungsgebiet binärer flüssiger Systeme. *Z. Phys. Chem.*, 57(1_2):56–73.

Hammond, B. and Stokes, R. (1955). Diffusion in binary liquid mixtures, Part 2. - The diffusion of carbon tetrachloride in come organic solvents at 25°. *Trans. Faraday Soc.*, 51:1641–1649.

Hardt, A. P., Anderson, D. K., Rathbun, R., Mar, B. W., and Babb, A. L. (1959). Self diffusion in liquids. II. Comparison between mutual and self-diffusion coefficients. *J. Phys. Chem.*, 63(12):2059–2061.

Harned, H. S. (1957). Ninth spiers memorial lecture. some recent experimental studies of diffusion in liquid systems. *Discuss. Faraday. Soc.*, 24:7–16.

Harris, K. R., Pua, C. K., and Dunlop, P. J. (1970). Mutual and tracer diffusion coefficients and frictional coefficients for the systems benzene–chlorobenzene, benzene–n-hexane, and benzene–n-heptane at 25°C. *J. Phys. Chem.*, 74:3518–3529.

Hatch, A., Kamholz, A. E., Hawkins, K. R., Munson, M. S., Schilling, E. A., Weigl, B. H., and Yager, P. (2001). A rapid diffusion immunoassay in a t-sensor. *Nat. Biotechnol.*, 19:461–465.

Häusler, E., Domagalski, P., Ottens, M., and Bardow, A. (2012). Microfluidic diffusion measurements: The optimal H-cell. *Chem. Eng. Sci.*, 72:45–50.

Heller, A., Giraudet, C., Makrodimitri, Z. A., Fleys, M. S. H., Chen, J., van der Laan, G. P., Economou, I. G., Rausch, M. H., and Fröba, A. P. (2016). Diffusivities of ternary mixtures of n-alkanes with dissolved gases by dynamic light scattering. *J. Phys. Chem. B*.

Heller, A., Koller, T. M., Rausch, M. H., Fleys, M. S. H., Bos, A. N. R., van der Laan, G. P., Makrodimitri, Z. A., Economou, I. G., and Fröba, A. P. (2014). Simultaneous determination of thermal and mutual diffusivity of binary mixtures of n-octacosane with carbon monoxide, hydrogen, and water by dynamic light scattering. *J. Phys. Chem. B*, 118(14):3981–3990.

Helmke, M. F., Simpkins, W. W., and Horton, R. (2004). Experimental determination of effective diffusion parameters in the matrix of fractured till. *Vadose Zone J.*, 3(3):1050–1056.

Hendriks, E., Kontogeorgis, G. M., Dohrn, R., Hemptinne, J.-C. d., Economou, I. G., Žilnik, L. F., and Vesovic, V. (2010). Industrial requirements for thermodynamics and transport properties. *Ind. Eng. Chem. Res.*, 49(22):11131–11141.

Hertz, H. and Mills, R. (1978). Velocity correlations in aqueous electrolyte solutions from diffusion, conductance, and transference data. Applications to concentrated solutions of 1:2 electrolytes. *J. Phys. Chem.*, 82(8):952–959.

Hertz, H. G., Harris, K. R., Mills, R., and Woolf, L. A. (1977). Velocity correlations in aqueous electrolyte solutions from diffusion, conductance, and transference data. Part 2, applications to concentrated solutions of 1–1 electrolytes. *Berich. Bunsen. Gesell.*, 81(7):664–670.

Hirschfelder, J. O. and Curtiss, C. F. (1949). The theory of flame propagation. *J. Chem. Phys.*, 17(11):1076–1081.

Hogervorst, W. (1971). Diffusion coefficients of noble-gas mixtures between 300 K and 1400 K. *Physica*, 51(1):59–76.

Holland-Letz, T., Dette, H., and Renard, D. (2012). Efficient algorithms for optimal designs with correlated observations in pharmacokinetics and dose-finding studies. *Biometrics*, 68(1):138–145.

Hopp, M., Mele, J., and Gross, J. (2018). Self-diffusion coefficients from entropy scaling using the pcp-saft equation of state. *Ind. Eng. Chem. Res.*, 57(38):12942–12950.

Houska, B., Telen, D., Logist, F., Diehl, M., and Van Impe, J. F. (2015). An economic objective for the optimal experiment design of nonlinear dynamic processes. *Automatica*, 51:98–103.

Jäger, B. and Bich, E. (2017). Thermophysical properties of krypton-helium gas mixtures from ab initio pair potentials. *J. Chem. Phys.*, 146(21):214302.

Jamali, S. H., Wolff, L., Becker, T. M., Bardow, A., Vlugt, T. J. H., and Moultos, O. A. (2018). Finite-size effects of binary mutual diffusion coefficients from molecular dynamics. *J. Chem. Theory Comp.*, 14(5):2667–2677.

Johnson, P. A. and Babb, A. L. (1956). Self-diffusion in liquids. 1. concentration dependence in ideal and non-ideal binary solutions. *J. Phys. Chem.*, 60(1):14–19.

Kalyakin, A., Volkov, A., Vylkov, A., Gorbova, E., Medvedev, D., Demin, A., and Tsiakaras, P. (2018). An electrochemical method for the determination of concentration and diffusion coefficient of ammonia-nitrogen gas mixtures. *J. Electroanal. Chem.*, 808:133–136.

Kamholz, A. E., Schilling, E. A., and Yager, P. (2001). Optical measurement of transverse molecular diffusion in a microchannel. *Biophys. J.*, 80:1967–1972.

Kamholz, A. E., Weigl, B. H., Finlayson, B. A., and Yager, P. (1999). Quantitative analysis of molecular interaction in a microfluidic channel: The T-sensor. *Anal. Chem.*, 71:5340–5347.

Käshammer, S., Weingärtner, H., and Hertz, H. G. (1994). Ternary diffusion in the system n-propanol + 1-chlorobutane + n-heptane at 25°C. *Z. Phys. Chem.*, 187:233–255.

Kestin, J., Knierim, K., Mason, E. A., Najafi, B., Ro, S. T., and Waldman, M. (1984). Equilibrium and transport properties of the noble gases and their mixtures at low density. *J. Phys. Chem. Ref. Data*, 13(1):229–303.

Kiefer, J. and Wolfowitz, J. (1960). The equivalence of two extremum problems. *Canadian J. Math.*, 12(363-366):234.

Klamt, A. (1995). Conductor-like screening model for real solvents: A new approach to the quantitative calculation of solvation phenomena. *J. Phys. Chem.*, 99(7):2224–2235.

Koller, T. M., Heller, A., Rausch, M. H., Wasserscheid, P., Economou, I.-G., and Fröba, A. P. (2015). Mutual and self-diffusivities in binary mixtures of [emim][b(cn)4] with dissolved gases by using dynamic light scattering and molecular dynamics simulations. *J. Phys. Chem. B*, 119(27):8583–8592.

Kontogeorgis, G. M. (2013). Association theories for complex thermodynamics. *Chem. Eng. Res. Des.*, 91(10):1840–1858.

Kooijman, H. A. and Taylor, R. (1991). Estimation of diffusion coefficients in multicomponent liquid systems. *Ind. Eng. Chem. Res.*, 30:1217–1222.

Kriesten, E., Voda, M., Bardow, A., Göke, V., Casanova, F., Blümich, B., Koß, H.-J., and Marquardt, W. (2009). Direct determination of the concentration dependence of diffusivities using combined model-based Raman and NMR experiments. *Fluid Phase Equilibr.*, 277:96–106.

Krishna, R. (2015). Serpentine diffusion trajectories and the ouzo effect in partially miscible ternary liquid mixtures. *Phys. Chem. Chem. Phys*, 17:27428–27436.

Krishna, R. and van Baten, J. M. (2005). The Darken relation for multicomponent diffusion in liquid mixtures of linear alkanes: An investigation using molecular dynamics (MD) simulations. *Ind. Eng. Chem. Res.*, 44:6939–6947.

Krüger, P., Schnell, S. K., Bedeaux, D., Kjelstrup, S., Vlugt, T. J. H., and Simon, J.-M. (2013). Kirkwood-Buff integrals for finite volumes. *J. Phys. Chem. Lett.*, 4(2):235–238.

Krüger, P. and Vlugt, T. J. H. (2018). Size and shape dependence of finite-volume kirkwood-buff integrals. *Phys. Rev. E*, 97:051301.

Kugler, T. (2015). *Determination of gaseous binary diffusion coefficients using a Loschmidt cell combined with holographic interferometry.* Dissertation, Universitat Erlangen-Nürnberg.

Kugler, T., Jäger, B., Bich, E., Rausch, M. H., and Fröba, A. P. (2013). Measurement of binary diffusion coefficients for neon–argon gas mixtures using a Loschmidt cell combined with holographic interferometry. *Int. J. Thermophys.*, 34(1):47–63.

Kugler, T., Jäger, B., Bich, E., Rausch, M. H., and Fröba, A. P. (2015a). Systematic study of mass transfer in a Loschmidt cell for binary gas mixtures. *Int. J. Thermophys.*, 36(10-11):3116–3132.

Kugler, T., Rausch, M., and Fröba, A. (2015b). Binary diffusion coefficient data of various gas systems determined using a loschmidt cell and holographic interferometry. *Int. J. Thermophys.*, 36(10-11):3169–3185.

Kullnick, M. (2001). *Interferometrische Untersuchung der Diffusion in binären Gemischen realer Gase mit einer Loschmidt-Diffusionsapparatur.* Dissertation, Technical University of Braunschweig.

Kuzmanović, B., van Delden, M. L., Kuipers, N. J. M., and de Haan, A. B. (2003). Fully automated workstation for liquid−liquid equilibrium measurements. *J. Chem. Eng. Data*, 48(5):1237–1244.

Landaverde-Cortes, D. C., Estrada-Baltazar, A., Iglesias-Silva, G. A., and Hall, K. R. (2007). Densities and viscosities of mtbe+ heptane or octane at p=0.1 MPa from (273.15 to 363.15) K. *J. Chem. Eng. Data*, 52(4):1226–1232.

Leaist, D. G. (1985). Moments analysis of restricted ternary diffusion: sodium sulfite + sodium hydroxide + water. *Can. J. Chemistry*, 63:2933–2939.

Leaist, D. G. (1990). Determination of ternary diffusion coefficients by the Taylor dispersion method. *J. Phys. Chem.*, 94:5180–5183.

Leffler, J. and Cullinan, H. T. J. (1970). Variation of liquid diffusion coefficients with composition. Binary systems. *Ind. Eng. Chem. Fund.*, 9:88–93.

Lefortier, S. G. R., Hamersma, P. J., Bardow, A., and Kreutzer, M. T. (2012). Rapid microfluidic screening of CO2 solubility and diffusion in pure and mixed solvents. *Lab Chip*, 12(18):3387–3391.

Leipertz, A. and Fröba, A. P. (2005). Diffusion measurements in fluids by dynamic light scattering. In *Diffusion in Condensed Matter*, pages 579–618. Springer.

Li, J., Doig, R., Camardese, J., Plucknett, K., and Dahn, J. R. (2015). Measurements of interdiffusion coefficients of transition metals in layered Li–Ni–Mn–Co oxide core–shell materials during sintering. *Chemistry of Materials*, 27(22):7765–7773.

Lin, Y., Yu, X., Wang, Z., Tu, S.-T., and Wang, Z. (2010). Measurement of temperature-dependent diffusion coefficients using a confocal raman microscope with microfluidic chips considering laser-induced heating effect. *Anal. Chim. Acta*, 667:103–112.

Lin, Y., Yu, X., Wang, Z., Tu, S.-T., and Wang, Z. (2012). Laminar flow diffusion interface control in a microchannel with accurate raman measurement. *Chem. Eng. Process.*, 57-58:1–7.

Liu, C., McGivern, W. S., Manion, J. A., and Wang, H. (2016). Theory and experiment of binary diffusion coefficient of n-alkanes in dilute gases. *J. Phys. Chem. A*, 120(41):8065–8074.

Liu, X., Bardow, A., and Vlugt, T. H. (2011a). Multicomponent maxwell-stefan diffusivities at infinite dilution. *Ind. Eng. Chem. Res.*, 50:4776–4782.

Liu, X., Martin-Calvo, A., McGarrity, E., Schnell, S. K., Calero, S., Simon, J.-M., Bedeaux, D., Kjelstrup, S., Bardow, A., and Vlugt, T. J. H. (2012). Fick diffusion coefficients in ternary liquid systems from equilibrium molecular dynamics simulations. *Ind. Eng. Chem. Res.*, 51:10247–10258.

Liu, X., Schnell, S. K., Simon, J.-M., Bedeaux, D., Kjelstrup, S., Bardow, A., and Vlugt, T. J. H. (2011b). Fick diffusion coefficients of liquid mixtures directly obtained from equilibrium molecular dynamics. *J. Phys. Chem. B*, 115:12921–12929.

Liu, X., Schnell, S. K., Simon, J.-M., Krüger, P., Bedeaux, D., Kjelstrup, S., Bardow, A., and Vlugt, T. J. H. (2013). Diffusion coefficients from molecular dynamics simulations in binary and ternary mixtures. *Int. J. Thermophys*, 34(7):1169–1196.

Liu, X., Vlugt, T. J., and Bardow, A. (2011c). Maxwell-Stefan diffusivities in liquid mixtures: Using molecular dynamics for testing model predictions. *Fluid Phase Equilibr.*, 301:110–117.

Liu, X., Vlugt, T. J. H., and Bardow, A. (2011d). Predictive Darken equation for Maxwell-Stefan diffusivities in multicomponent mixtures. *Ind. Eng. Chem. Res.*, 50:10350–10358.

Longsworth, L. (1966). The diffusion of hydrogen bonding solutes in carbon tetrachloride. *J. Colloid. Interf. Sci.*, 22(1):3–11.

MacQuigg, D. R. (1977). Hologram fringe stabilization method. *Appl. Opt.*, 16(2):291–292.

Makrodimitri, Z. A., Heller, A., Koller, T. M., Rausch, M. H., Fleys, M. S. H., Bos, A. R., van der Laan, G. P., Fröba, A. P., and Economou, I. G. (2015). Viscosity of heavy n -alkanes and diffusion of gases therein based on molecular dynamics simulations and empirical correlations. *J. Chem. Thermodyn.*, 91:101–107.

Makrodimitri, Z. A., Unruh, D. J. M., and Economou, I. G. (2011). Molecular simulation of diffusion of hydrogen, carbon monoxide, and water in heavy n-alkanes. *J. Phys. Chem. B*, 115:1429–1439.

Marbach, W., Hertz, H. G., and Weingärtner, H. (1995). Self- and mutual diffusion coefficients of some binary liquid n-alkane mixtures – a velocity correlation study –. *Z. Phys. Chem.*, 189(1):63–79.

Marine, N. A., Klein, S. A., and Posner, J. D. (2009). Partition coefficient measurements in picoliter drops using a segmented flow microfluidic device. *Anal. Chem.*, 81(4):1471–1476.

Marquardt, W. (2005). Model-based experimental analysis of kinetic phenomena in multi-phase reactive systems. *Chem. Eng. Res. Des.*, 83(6):561–573.

Marrero, T. R. and Mason, E. A. (1972). Gaseous diffusion coefficients. *J. Phys. Chem. Ref. Data*, 1(1):3–118.

Martinson, W. S. and Barton, P. I. (2000). A differentiation index for partial differential-algebraic equations. *SIAM J. Sci. Comput.*, 21(6):2295–2315.

Masuch, K. (2011). *Eindimensionale Ramanspektroskopie zur Temperatur- und Konzentrationsmessung in Hydrogelen.* Dissertation, RWTH Aachen University.

Matos Lopes, M. L. S., Nieto de Castro, C. A., and Sengers, J. V. (1992). Mutual diffusivity of a mixture of n-hexane and nitrobenzene near its consolute point. *Int. J. Thermophys.*, 13(2):283–294.

McCall, D. W. and Douglass, D. C. (1967). Diffusion in binary solutions. *J. Chem. Phys.*, 71(4):987–997.

McCarty, K. P. and Mason, E. A. (1960). Kirkendall effect in gaseous diffusion. *Phys. Fluids.*, 3(6):908–922.

Medina, J. L. and Ramírez, C. A. (2016). Theoretical and experimental estimation of binary gas diffusivities in a nonisothermal Stefan diffusion column. *Chem. Eng. Commun.*, 203(12):1625–1640.

Mialdun, A., Sechenyh, V., Legros, J. C., Ortiz de Zarate, J. M., and Shevtsova, V. (2013). Investigation of Fickian diffusion in the ternary mixture of 1,2,3,4-tetrahydronaphthalene, isobutylbenzene, and dodecane. *J. Chem. Phys.*, 139:104903.

Michelsen, M. L. and Mollerup, J. (2007). *Thermodynamic Models - Fundamentals and Computational Aspects.* Tie-Line Publications, Holte.

Miller, D. G., Sartorio, R., Paduano, L., Rard, J. A., and Albright, J. G. (1996). Effects of different sized concentration differences across free diffusion boundaries and comparison of Gouy and Rayleigh diffusion measurements using NaCl-KCl-H2O. *J. Solution Chem.*, 25(12):1185–1211.

Miller, L. and Carman, P. C. (1961). Self-diffusion in mixtures. Part 4.– Comparison of theory and experiment for certain gas mixtures. *Trans. Faraday Soc.*, 57(0):2143–2150.

Miller, T., Lamb, B., Prater, K., Lee, J., and Adams, R. (1964). Tracer diffusion coefficients of aromatic organic molecules. *Anal. Chem.*, 36(2):418–420.

Mills, R. (1965). The intradiffusion and derived frictional coefficients for benzene and cyclohexane in their mixtures at 25°. *J. Phys. Chem.*, 69(9):3116–3119.

Mills, R. and Hertz, H. G. (1980). Application of the velocity cross-correlation method to binary nonelectrolyte mixtures. *J. Phys. Chem.*, 84(2):220–224.

Mills, R. and Woolf, L. A. (1992). Velocity cross-correlation coefficients for the binary systems methanol-water and acetone-water at 278 K and 298 K. *J. Mol. Liq.*, 52:115–130.

Milzetti, J., Nayar, D., and van der Vegt, N. F. A. (2018). Convergence of kirkwood-buff integrals of ideal and nonideal aqueous solutions using molecular dynamics simulations. *J. Phys. Chem. B*, 122(21):5515–5526.

Mitsos, A., Bollas, G. M., and Barton, P. I. (2009). Bilevel optimization formulation for parameter estimation in liquid–liquid phase equilibrium problems. *Chem. Eng. Sci.*, 64(3):548–559.

Moggridge, G. (2012a). Prediction of the mutual diffusivity in binary liquid mixtures containing one dimerising species, from the tracer diffusion coefficients. *Chem. Eng. Sci.*, 76:199–205.

Moggridge, G. D. (2012b). Prediction of the mutual diffusivity in binary non-ideal liquid mixtures from the tracer diffusion coefficients. *Chem. Eng. Sci.*, 71:226–238.

Moultos, O. A., Tsimpanogiannis, I. N., Panagiotopoulos, A. Z., and Economou, I. G. (2014). Atomistic molecular dynamics simulations of CO2 diffusivity in H2O for a wide range of temperatures and pressures. *J. Phys. Chem. B*, 118(20):5532–5541.

Moultos, O. A., Tsimpanogiannis, I. N., Panagiotopoulos, A. Z., Trusler, J. P. M., and Economou, I. G. (2016). Atomistic molecular dynamics simulations of carbon dioxide diffusivity in n-hexane, n-decane, n-hexadecane, cyclohexane, and squalane. *J. Phys. Chem. B*, 120(50):12890–12900.

Munson, M. S., Hawkins, K. R., Hasenbank, M. S., and Yager, P. (2005). Diffusion based analysis in a sheath flow microchannel: the sheath flow t-sensor. *Lab Chip*, 5:856–862.

Mutoru, J. W. and Firoozabadi, A. (2011). Form of multicomponent Fickian diffusion coefficients matrix. *J. Chem. Thermodyn.*, 43:1192–1203.

Neumann, D. B. and Rose, H. W. (1967). Improvement of recorded holographic fringes by feedback control. *Appl. Opt.*, 6(6):1097–1104.

NIST (2018). http://webbook. nist. gov.

Paccagnella, A., Ottaviani, G., Fabbri, P., Ferla, G., and Queirolo, G. (1985). Silicon diffusion in aluminium. *Thin Solid Films*, 128(3-4):217–223.

Pappaert, K., Biesemans, J., Clicq, D., Vankrunkelsven, S., and Desmet, G. (2005). Measurements of diffusion coefficients in 1-D micro- and nanochannels using shear-driven flows. *Lab Chip*, 5:1104–1110.

Parez, S., Guevara-Carrion, G., Hasse, H., and Vrabec, J. (2013). Mutual diffusion in the ternary mixture of water + methanol + ethanol and its binary subsystems. *Phys. Chem. Chem. Phys.*, 15:3985–4001.

Pelletier, M. J. (2003). Quantitative analysis using raman spectrometry. *Appl. Spectrosc.*, 57(1):20A–42A.

Peters, C., Thien, J., Wolff, L., Koß, H.-J., and Bardow, A. (2019). Quaternary diffusion coefficients in liquids from microfluidics and raman microspectroscopy: Cyclohexane + toluene + acetone + methanol. *J. Chem. Eng. Data*, xx(xx):xx.

Peters, C., Wolff, L., Haase, S., Thien, J., Brands, T., Koß, H.-J., and Bardow, A. (2017). Multicomponent diffusion coefficients from microfluidics using Raman microspectroscopy. *Lab Chip*, 17:2768–2776.

Peters, C., Wolff, L., Vlugt, T. J. H., and Bardow, A. (2016). Chapter 5 diffusion in liquids: Experiments, molecular dynamics, and engineering models. In *Experimental Thermodynamics Volume X: Non-equilibrium Thermodynamics with Applications*, pages 78–104. The Royal Society of Chemistry.

Poling, B. E., Prausnitz, J. M., and O'Connel, J. P. (2001). *The Properties of Gases and Liquids*, volume 5. McGraw-Hill, New York.

Prabhakar, S. and Weingärtner, H. (1983). The influence of molecular association on diffusion in the system methanol–carbon tetrachloride at 25°C. *Z. Phys. Chem.*, 137(1):1–12.

Price, W. E. (1988). Theory of the Taylor dispersion technique for three-component-system diffusion measurements. *J. Chem. Soc. Faraday Trans.I*, 84:2431–2439.

Pronzato, L. and Walter, E. (1988). Robust experiment design via maximin optimization. *Math. Biosci.*, 89(2):161–176.

Rapaport, D. C. (2004). *The Art of Molecular Dynamics Simulation.* Cambridge University Press.

Rathbun, R. E. and Babb, A. L. (1966). Empirical method for prediction of concentration dependence of mutual diffusivities in binary mixtures of associated and nonpolar liquids. *Ind. Eng. Chem. Proc. Dd.*, 5(3):273–275.

Recker, S., Kerimoglu, N., Harwardt, A., Tkacheva, O., and Marquardt, W. (2013). On the integration of model identification and process optimization. *Comput.-Aided Chem. En.*, 32:1021–1026.

Recker, S., Kühl, P., Diehl, M., Bock, H. G., and Marquardt, W. (2012). Sigmapoint approach for robust optimization of nonlinear dynamic systems. *SIMULTECH*, pages 199–207.

Rehfeldt, S. and Stichlmair, J. (2007). Measurement and calculation of multicomponent diffusion coefficients in liquids. *Fluid Phase Equilibr.*, 256:99–104.

Rehfeldt, S. and Stichlmair, J. (2010). Measurement and prediction of multicomponent diffusion coefficients in four ternary liquid systems. *Fluid Phase Equilibr.*, 290:1–14.

Renon, H. and Prausnitz, J. M. (1968). Local compositions in thermodynamic excess functions for liquid mixtures. *AIChE J.*, 14(1):135–144.

Rohsenow, W., Hartnett, J. P., and Cho, Y. I. (1998). *Handbook of Heat Transfer*. McGraw-Hill Professional, 3rd edition.

Ryckeboer, E., Vierendeels, J., Lee, A., Werquin, S., Bienstman, P., and Baets, R. (2013). Measurement of small molecule diffusion with an optofluidic silicon chip. *Lab Chip*, 13:4392–4399.

Salmon, J.-B., Ajdari, A., Tabeling, P., Servant, L., Talaga, D., and Joanicot, M. (2005). In situ Raman imaging of interdiffusion in a microchannel. *Appl. Phys. Lett.*, 86:094106.

Sanchez, V. and Clifton, M. (1977). An empirical relationship for predicting the variation with concentration of diffusion coefficients in binary liquid mixtures. *Ind. Eng. Chem. Fundam.*, 16(3):318–320.

Sanni, S. A., Fell, C. J. D., and Hutchison, H. P. (1971). Diffusion coefficients and densities for binary organic liquid mixtures. *J. Chem. Eng. Data*, 16:424–427.

Schöneberger, J. C., Arellano-Garcia, H., and Wozny, G. (2010). Local optima in model-based optimal experimental design. *Ind. Eng. Chem. Res.*, 49(20):10059–10073.

Shapiro, A. A. (2003). Evaluation of diffusion coefficients in multicomponent mixtures by means of the fluctuaction theory. *Physica A*, 320:211–234.

Sharipov, F. and Benites, V. J. (2017). Transport coefficients of helium-neon mixtures at low density computed from ab initio potentials. *J. Chem. Phys.*, 147(22):224302.

Shevtsova, V., Santos, C., Sechenyh, V., Legros, J. C., and Mialdun, A. (2014). Diffusion and Soret in ternary mixtures. Preparation of the DCMIX2 experiment on the ISS. *Microgravity Sci. Tec.*, 25:275–283.

Silvey, S., Titterington, D., and Torsney, B. (1978). An algorithm for optimal designs on a design space. *Commun. Stat. A-Theor.*, 7(14):1379–1389.

Spiess, A. C., Zavrel, M., Ansorge-Schumacher, M. B., Janzen, C., Michalik, C., Schmidt, T. W., Schwendt, T., Büchs, J., Poprawe, R., and Marquardt, W. (2008). Model discrimination for the propionic acid diffusion into hydrogel beads using lifetime confocal laser scanning microscopy. *Chem. Eng. Sci.*, 63(13):3457–3465.

Srivastava, B. N. and Paul, R. (1962). Multicomponent diffusion in the system 85Kr-He-Kr. *Physica*, 28(7):646–652.

Staker, G. R. and Dunlop, P. J. (1976). The pressure dependence of the mutual diffusion coefficients of binary mixtures of helium and six other gases at 300 K: tests of Thorne's equation. *Chem. Phys. Lett.*, 42(3):419–422.

Staker, G. R., Dunlop, P. J., Harris, K. R., and Bell, T. N. (1975). The pressure and composition dependence of mutual diffusion in the system helium-nitrogen at 300 K. *Chem. Phys. Lett.*, 32(3):561–565.

Staker, G. R., Yabsley, M. A., Symons, J. M., and Dunlop, P. J. (1974). Concentration dependence of the mutual diffusion coefficients of the systems He+Ne, He+Ar, He+Kr, He+Xe and Ne+Ar at 300 K and one atmosphere. evaluation of potential parameters from isothermal measurements. *J. Chem. Soc. Farad. T. 1*, 70:825–831.

Stewart, W. E., Gotoh, S., and Sørensen, J. P. (1973). Studies of the Loschmidt diffusion experiment. I. A perturbation analysis of the diffusion cell. *Ind. Eng. Chem. Fund.*, 12(1):114–118.

Strigle, R. (1994). *Packed Tower Design and Applications: Random and Structured Packings.* Gulf Publishing Company.

Suárez, I. J., Borrás, C., Scharifker, B. R., and Mostany, J. (2007). Chapter 11 diffusion in solids: Hydrogen transport in massive and microdispersed palladium. In *Electrochemistry and Materials Engineering*, pages 173–193. Research Signpost.

Suarez-Iglesias, O., Medina, I., Pizarro, C., and Bueno, J. L. (2007). On predicting self-diffusion coefficients from viscosity in gases and liquids. *Chem. Eng. Sci.*, 62:6499–6515.

Suarez-Iglesias, O., Medina, I., Sanz, M. d. l. A., Pizarro, C., and Bueno, J. L. (2015). Self-diffusion in molecular fluids and noble gases: Available data. *J. Chem. Eng. Data*, 60(10):2757–2817.

Tanaka, S., Clewley, J. D., and Flanagan, T. B. (1977). Kinetics of hydrogen absorption by lanthanum-nickel (lani5). *J. Phys. Chem.*, 81(17):1684–1688.

Taylor, R. and Krishna, R. (1993). *Multicomponent Mass Transfer*. John Wiley & Sons, INC.

Telen, D., Logist, F., Van Derlinden, E., and Van Impe, J. (2013). On the trade-off between experimental effort and information content in optimal experimental design for calibrating a predictive microbiology model. *Journal de la Societe Francaise de Statistique & Revue de Statistique Appliquee*, 154(3):95–112.

Telen, D., Vallerio, M., Cabianca, L., Houska, B., Van Impe, J., and Logist, F. (2015). Approximate robust optimization of nonlinear systems under parametric uncertainty and process noise. *J. Process Contr.*, 33:140–154.

Telen, D., Vercammen, D., Logist, F., and Van Impe, J. (2014). Robustifying optimal experiment design for nonlinear, dynamic (bio)chemical systems. *Comput. Chem. Eng.*, 71:415–425.

Thien, J., Peters, C., Brands, T., Koß, H.-J., and Bardow, A. (2017). Efficient determination of liquid-liquid equilibria using microfluidics and raman microspectroscopy. *Ind. Eng. Chem. Res.*, 56(46):13905–13910.

Thien, J., Reinpold, L., Brands, T., Koß, H.-J., and Bardow, A. (2019). Automated physical property measurements from calibration to data analysis: Microfluidic platform for liquid-liquid equilibrium using raman microspectroscopy. *J. of Chem. Eng. Data*, xx(xx):xx.

Titterington, D. (1976). Algorithms for computing D-optimal designs on a finite design space. In *Proc. of the 1976 Conf. on Information Science and Systems, John Hopkins University*, volume 3, pages 213–216.

Tuckerman, M. E. (2010). *Statistical Mechanics: Theory and Molecular Simulation*. Oxford University Press, Oxford.

Tyn, M. T. and Calus, W. F. (1975). Temperature and concentration dependence of mutual diffusion coefficients of some binary liquid systems. *J. Chem. Eng. Data*, 20:310–316.

van de Ven-Lucassen, I. M. J. J., Kemmere, M. F., and Kerhof, P. J. A. M. (1997). Complications in the use of the Taylor dispersion method for ternary diffusion measurements: Methanol + acetone + water mixtures. *J. Solution Chem.*, 26(12):1145–1167.

van Heijningen, R. J. J., Harpe, J. P., and Beenakker, J. J. M. (1968). Determination of the diffusion coefficients of binary mixtures of the noble gases as a function of temperature and concentration. *Physica*, 38(1):1–34.

Vercher, E., Orchillés, A. V., Miguel, P. J., and Martínez-Andreu, A. (2007). Volumetric and ultrasonic studies of 1-ethyl-3-methylimidazolium trifluoromethanesulfonate ionic liquid with methanol, ethanol, 1-propanol, and water at several temperatures. *J. Chem. Eng. Data*, 52(4):1468–1482.

Vignes, A. (1966). Diffusion in binary solutions. Variation of diffusion coefficient with composition. *Ind. Eng. Chem. Fund.*, 5:189–199.

Walter, E. and Pronzato, L. (1987). Optimal experiment design for nonlinear models subject to large prior uncertainties. *Am. J. Physiol.*, 253(3):R530–R534.

Walter, E. and Pronzato, L. (1997). *Identification of Parametric Models.* Masson.

Walz, O., Djelassi, H., and Mitsos, A. (2020). Optimal experimental design for optimal process design: A trilevel optimization formulation. *AIChE J.*, 66(1):e16788.

Wang, J. (1951). Self-diffusion and structure of liquid water. I. measurement of self-diffusion of liquid water with deuterium as tracer. *J. Am. Chem. Soc.*, 73(2):510–513.

Wang, Y., Lin, Q., and Mukherjee, T. (2005). A model for laminar diffusion-based complex electrokinetic passive micromixers. *Lab Chip*, 5:877–887.

Weingärtner, H. (1990). The microscopic basis of self diffusion – mutual diffusion relationships in binary liquid mixtures. *Ber. Bunsenges. Phys. Chem.*, 94:358–364.

Weingärtner, H. (2005). The molecular description of mutual diffusion processes in liquid mixtures. In *Diffusion in Condensed Matter*, pages 555–578. Springer.

Weingärtner, H. and Bertagnolli, H. (1986). The microscopic description of mutual diffusion and closely related transport processes in liquid mixtures. *Ber. Bunsenges. Phys. Chem.*, 90(12):1167–1174.

Werts, M. H. V., Raimbault, V., Texier-Picard, R., Poizat, R., Français, O., Griscom, L., and Navarro, J. R. G. (2012). Quantitative full-colour transmitted light microscopy and dyes for concentration mapping and measurement of diffusion coefficients in microfluidic architectures. *Lab Chip*, 12:808–820.

Whitesides, G. M. (2006). The origins and the future of microfluidics. *Nature*, 442:368–373.

Winkelmann, J. (2007). *Diffusion in Gases, Liquids and Electrolytes A: Gases in gases, liquids and their mixtures.* Springer Materials.

Wolff, L., Jamali, S. H., Becker, T. M., Moultos, O. A., Vlugt, T. J. H., and Bardow, A. (2018a). Prediction of composition-dependent self-diffusion coefficients in binary liquid mixtures: The missing link for Darken-based models. *Ind. Eng. Chem. Res.*, 57(43):14784–14794.

Wolff, L., Koß, H.-J., and Bardow, A. (2016). The optimal diffusion experiment. *Chem. Eng. Sci.*, 152:392–402.

Wolff, L., Zangi, P., Brands, T., Rausch, M. H., Koß, H.-J., Fröba, A. P., and Bardow, A. (2018b). Concentration-dependent diffusion coefficients of binary gas mixtures using a loschmidt cell with holographic interferometry, Part i: Multiple experiments. *Int. J. Thermophys.*, 39(12):133.

Wolff, L., Zangi, P., Brands, T., Rausch, M. H., Koß, H.-J., Fröba, A. P., and Bardow, A. (2018c). Concentration-dependent diffusion coefficients of binary gas mixtures using a loschmidt cell with holographic interferometry, Part ii: Single experiment. *Int. J. Thermophys.*, 39(12):132.

Woolf, L. A., Mills, R., Leaist, D. G., Erkey, C., Akgerman, A., Easteal, A. J., Miller, D. G., Albright, J. G., Li, S. F. Y., and Wakeham, W. (1991). *Measurement of the Transport Properties of Fluids*, chapter 9 - Diffusion Coefficients, pages 227–320. Blackwell Scientific Publications, Oxford.

Wright, J. E., Stevens, G. W., Kelly, E. D., and White, L. R. (1994). Measurement of diffusion coefficient using a closed capillary technique. *AIChE J.*, 40(2):365–368.

Wu, G., Fiebig, M., and Leipertz, A. (1988). Messung des binären Diffusionskoeffizienten in einem Entmischungssystem mit Hilfe der Photonen-Korrelationsspektroskopie. *Warme Stoffubertrag.*, 22(6):365–371.

Yoshinobu, K. and Yasumichi, O. (1972). Self-diffusion coefficients and interdiffusion coefficient in acetone–benzene system. *B. Chem. Soc. Jpn.*, 45(8):2437–2439.

Yu, Y. et al. (2010). Monotonic convergence of a general algorithm for computing optimal designs. *Ann. Stat.*, 38(3):1593–1606.

Yue, F., Fu, P., Liu, Y., Bie, K., and Zhou, H. (2018). Determination of the binary gas diffusion coefficients using large lateral shearing interferometry. *Optik*, 156:825–833.

Zhang, Y. (2010). Diffusion in minerals and melts. *Reviews in Mineralogy and Geochemistry*, 72(1):1–1038.

Zhou, M., Yuan, X. G., Zhang, Y., and Yu, K. T. (2013). A local composition based Maxwell-Stefan diffusivity model for binary liquid systems. *Ind. Eng. Chem. Res.*, 52(31):10845–10852.

Zhu, Q., Moggridge, G. D., and D'Agostino, C. (2015). A local composition model for the prediction of mutual diffusion coefficients in binary liquid mixtures from tracer diffusion coefficients. *Chem. Eng. Sci.*, 132:250–258.

Aachener Beiträge zur Technischen Thermodynamik

ABTT 1
Philip Voll
Automated Optimization-Based Synthesis of Distributed Energy Supply Systems
1. Auflage 2014
ISBN 978-3-86130-474-6

ABTT 2
Johannes Jung
Comparative Life Cycle Assessment of Industrial Multi-Product Processes
1. Auflage 2014
ISBN 978-3-86130-471-5

ABTT 3
Franz Lanzerath
Modellgestützte Entwicklung von Adsorptionswärmepumpen
1. Auflage 2014
ISBN 978-3-86130-472-2

ABTT 4
Thorsten Brands
Einfluss der Gemischzusammensetzung auf die Verbrennung im Diesel- und GCAI-Motor
1. Auflage 2014
ISBN 978-3-95886-006-3

ABTT 5
Dominique Dechambre
Efficient Measurement of Liquid-Liquid Equilibria using Automation and Optimal Experimental Design
1. Auflage 2016
ISBN 978-395886-077-3

ABTT 6
Niklas von der Aßen
From Life-Cycle Assesement towards life-Cycle Design of Carbon Dioxide Capture and Utilization
1. Auflage 2016
ISBN 978-3-95886-080-3

ABTT 7
Matthias Lampe
Integrated Process and Organic Rankine Cycle Working Fluid Design in the Continuous-Molecular Targeting Framework
1. Auflage 2016
ISBN 978-3-95886-086-5

ABTT 8
Thomas Hülser
Optische Untersuchung der Zündvorgänge und deren Auswirkung auf die Verbrennung in PKW-Motoren
1. Auflage 2016
ISBN 978-3-95886-090-2

Aachener Beiträge zur Technischen Thermodynamik

ABTT 9
Malte Döntgen
Reaction Models from Reactive Molecular Dynamics and High-Level Kinetics Predictions
1. Auflage 2016
ISBN 978-3-95886-156-5

ABTT 10
Heike Schreiber
Experiments and Validated Models for Adsorption Thermal Energy Storage in Industrial and Residential Application
1. Auflage 2017
ISBN 978-3-95886-178-7

ABTT 11
André Dirk Sternberg
System-Wide Perspective for Life Cycle Assesment of CO_2-based C1-Chemicals
1. Auflage 2017
ISBN 978-3-95886-193-0

ABTT 12
Uwe Bau
From Dynamic Simulation to Optimal Design and Control of Adsorption Energy Systems
1. Auflage 2018
ISBN 978-3-95886-216-6

ABTT 13
Christian Jens
Modellbasiertes Design von Produkt, Lösungsmittel und Prozess für die Ameisensäure-synthese aus CO_2 und H_2
1. Auflage 2018
ISBN 978-3-95886-231-9

ABTT 14
Jan David Scheffczyk
Integrated Computer-Aided Design of Molecules and Processes using COSMO-RS
1. Auflage 2018
ISBN 978-3-95886-236-4

ABTT 15
Björn Bahl
Optimization-Based Synthesis of Large-Scale Energy Systems by Time-Series Aggregation
1. Auflage 2018
ISBN 978-3-95886-240-1

Aachener Beiträge zur Technischen Thermodynamik

ABTT 16
Bastian Liebergesell
A Milliliter-Scale Setup for the Efficient Characterization of Multicomponent Vapor-Liquid Equilibria Using Raman Spectroscopy
1. Auflage 2018
ISBN 978-3-95886-247-0

ABTT 17
Stefan Wilhelm Graf
A Design Approach for Adsorption Energy Systems Integrating Dynamic Modeling with Small-Scale Experiments
1. Auflage 2018
ISBN 978-3-95886-258-6

ABTT 18
Sebastian Kaminski
Quantum-Mechanics-Based Prediction of SAFT Parameters for Non-Associating and Associating Molecules Containing Carbon, Hydrogen, Oxygen and Nitrogen
1. Auflage 2019
ISBN 978-3-95886-270-8

ABTT 19
Maike Renate Hennen
Decision Support for the Synthesis of Energy Systems by Analysis of the Near-Optimal Solution Space
1. Auflage 2019
ISBN 978-3-95886-277-7

ABTT 20
Peyman Yamin
COSMO-RS-Based Methods for Improved Modelling of Complex Chemical Systems
1. Auflage 2019
ISBN 978-3-95886-288-3

ABTT 21
Meltem Erdogan
Assessement of Adsorbents for Drying by Experiments and Dynamic Simulations
1. Auflage 2019
ISBN 978-3-95886-303-3

ABTT 22
Christian Schulz
SRS/LIF-Messungen zur Charakterisierung rußarmer dieselähnlicher Flammen von alternativon Kraftotoffon und n Hoptan
1. Auflage 2019
ISBN 978-3-91886-310-1

Aachener Beiträge zur Technischen Thermodynamik

ABTT 23
Peter Beumers
Physically-Based Models for the Analysis of Raman Spectra
1. Auflage 2019
ISBN 978-3-95886-319-4

ABTT 24
Arne Kätelhön
Technology Choice Model for Consequential Life Cycle Assessment
1. Auflage 2019
ISBN 978-3-95886-324-8

ABTT 25
Christine Peters
Measurement of Multicomponent Diffusion in Liquids Using Raman Microspectroscopy and Microfluidics
1. Auflage 2020
ISBN 978-3-95886-337-8

ABTT 26
Dinah Elena Hollermann
Reliable and Robust Optimal Design of Sustainable Energy Systems
1. Auflage 2020
ISBN 978-3-95886-346-0

ABTT 27
Thomas Raffius
Laserspektroskopische Analyse von selbstzündenden motorischen Einspritzstrahlen alternativer Biokraftstoffe
1. Auflage 2020
ISBN 978-3-95886-358-3

ABTT 28
Johannes Schilling
Integrated Thermo-Economic Design of Processes and Molecules Using PC-SAFT
1. Auflage 2020
ISBN 978-3-95886-368-2

ABTT 29
Nils Julius Baumgärtner
Optimization of Low-Carbon Energy Systems from Industrial to National Scale
1. Auflage 2020
ISBN 978-3-95886-385-9

ABTT 30
Ludger Wolff
From Model-based Experimental Design and Analysis of Diffusion and Liquid-Liquid Equilibria to Process Applications
1. Auflage 2021
ISBN 978-3-95886-402-3